Progress in Mathematics

4   GENERAL

Progress in Mathematics

Pupils' Book 1G
1G Mental Tests and Phase Tests
1G Copy Masters
Pupils' Book 2G
2G Mental Tests and Phase Tests
2G Copy Masters
Pupils' Book 3G
3G Mental Tests and Phase Tests
3G Copy Masters
Pupils' Book 4G
4G Mental Tests and Phase Tests
4G Copy Masters

Pupils's Book 1C
1C Mental Tests and Phase Tests
1C Copy Masters
Pupils' Book 2C
2C Mental Tests and Phase Tests
2C Copy Masters
Pupils' Book 3C
3C Mental Tests and Phase Tests
3C Copy Masters
Pupils' Book 4C
4C Mental Tests and Phase Tests
4C Copy Masters

5th-year volumes are in preparation, as is a third 'layer'—the 'E' ('extension') books.

Progress in Mathematics

4   GENERAL

Les Murray BA
Senior Teacher and Head of Mathematics, Garstang County High School

Stanley Thornes (Publishers) Ltd

© Les Murray 1986

First published in 1986 by Stanley Thornes (Publishers) Ltd, Old Station Drive, Leckhampton, Cheltenham GL53 0DN, UK

British Library Cataloguing in Publication Data

Murray, Les.
    Progress in mathematics.
    4G
    1. Mathematics—Examinations, questions, etc.
    I. Title
    510'.76   QA43

    ISBN 0-85950-178-7

Typeset by KEYTEC, Bridport, Dorset.
Printed and bound in Great Britain at The Bath Press, Avon.

Preface

This book has been written with the middle level of the GCSE examination in mind following a detailed analysis of all the available examination syllabuses.

The book contains numerous, carefully graded questions. It is not intended to be worked through from cover to cover; the teacher should be selective in the use of exercises and questions.

Revision sections have been inserted at intervals, questions being based on previous chapters. There is one revision exercise per chapter thus allowing one topic to be revised at a time. At the end of the book, twelve miscellaneous revision papers have been provided. The first eight of these papers each revise a limited number of topics (either seven or eight chapters each). Revision Paper 9 covers the whole of the book, while Revision Papers 10 to 12 are a miscellaneous selection including work covered in earlier books. This format allows pupils to be introduced gently to an examination situation through the steady progression from one topic to papers containing miscellaneous topics. The course continues to include calculator work, investigations and open-ended questions providing an opportunity for further study.

Photocopy masters are again available to the teacher for exercises where pupils may benefit by their provision. Such exercises have been labelled **M**.

The completion of this book has been dependent on the valued help and advice given to me by many people, in particular Mr Roger Wilson, Head of Mathematics at Parklands High School, Chorley, who has carefully and painstakingly worked through the whole text and has provided the answers as well as giving welcome advice; and to Mr J. Britton, Head of Mathematics at Copthall School, London, for his most useful comments. My thanks also go to staff and pupils of Garstang County High School, for their interest and co-operation while writing has been in progress; to Casio Electronics and Texas Instruments for the loan of a selection of calculators thus enabling me to consider the different characteristics of calculators in my writing; to CCM Ltd, Garstang and the numerous people who have supplied invaluable information.

Les Murray
1986

To RLM

Acknowledgements

The author and publishers are grateful to the following:

British Railways Board for the timetables on p.47 and p.48.

British Telecom for the table on p.44.

Club Cantabrica for the price list on p.137.

Van Nostrand Reinhold Co Ltd for the Chinese version of Pascal's triangle on p.36.

Contents

1 Sets

1. From set T, list:
(a) the odd numbers,
(b) the factors of 12,
(c) the multiples of 4,
(d) the prime numbers,
(e) the numbers that are multiples of both 2 and 3.

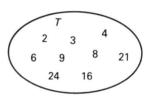

2. In the Venn diagram,
R = the set of red cars and
E = the set of estate cars.
Copy the diagram and shade the region that shows the set of red estate cars.

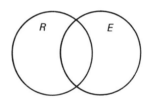

3. In the Venn diagram, $\mathscr{E}$ = {*natural numbers less than 21}. Make a copy of the diagram and write each of the following numbers in the correct part (each number should be written only once).

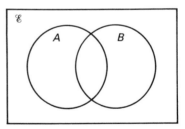

(a) 10 is in both A and B.
(b) 5 is in A but not in B.
(c) 2 is in B but not in A.
(d) 3 is neither in A nor in B.
(e) 4 is in B but not in A.
(f) 19 is neither in A nor in B.
(g) 16 is in B but not in A.
(h) 1 is not in A or in B.
(i) 20 is in both A and B.
(j) 17 is not in A or in B.
(k) 14 is in B but not in A.
(l) 9 is neither in A nor in B.

*See the glossary, p.478.

1

(m) 15 is in A but not in B.   (q) 7 is neither in A nor in B.
(n) 6 is in B but not in A.   (r) 12 is in B but not in A.
(o) 11 is not in A or in B.   (s) 13 is neither in A nor in B.
(p) 8 is in B but not in A.   (t) 18 is in B but not in A.

4. In the diagram produced in question 3 describe:
(a) set A (b) set B

Exercise 2

For each question, list the numbers or letters that are:
(a) in both P and Q,   (b) in P or Q or both.

1. $P = \{4, 5, 6, 7\}$ $Q = \{6, 7, 8\}$
2. $P = \{R, L, M\}$ $Q = \{J, G, M\}$
3. $P = \{2, 5, 7\}$ $Q = \{7, 6, 5\}$
4. $P = \{2, 8, 6, 12\}$ $Q = \{2, 10, 12, 6\}$
5. $P = \{4, 9, 1\}$ $Q = \{1, 4, 9\}$
6. $P = \{5, 10, 15, 20\}$ $Q = \{12, 16, 20\}$
7. $P = \{0, 1, 2, 3\}$ $Q = \{1, 2, 3, 4\}$
8. $P = \{A, T\}$ $Q = \{A, N\}$
9. $P = \{a, e, i, o, u\}$ $Q = \{f, a, c, e\}$
10. $P = \{10, 9, 8, 7\}$ $Q = \{7, 8, 9, 10\}$

Exercise 3

Answer these questions using the given Venn diagrams:

1. $\mathscr{E} = \{$natural numbers less than 13$\}$
$F = \{$factors of 24$\}$
$V = \{$even numbers$\}$

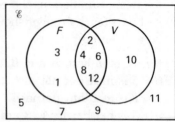

(a) Which numbers are both even and factors of 24?
(b) Which number is even but is not a factor of 24?
(c) Which numbers are not factors of 24?
(d) Which numbers are neither even nor factors of 24?

2

2. $\mathscr{E}$ = {the first 14 letters in the alphabet}

F = {letters in the name Fibonacci}

V = {vowels}

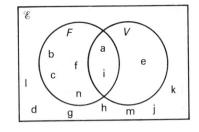

(a) Which vowels are in set F?

(b) Which letters are neither vowels nor in the name Fibonacci?

3. $\mathscr{E}$ = {even numbers < 32}

S = {multiples of 6}

F = {multiples of 4}

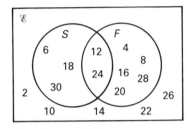

(a) Which numbers are multiples of both 4 and 6?

(b) Which numbers are multiples of 6 but not multiples of 4?

(c) Which numbers are neither multiples of 6 nor of 4?

(d) Which numbers are either multiples of 6, multiples of 4, or multiples of both 6 and 4?

4. (a) Copy the Venn diagram:

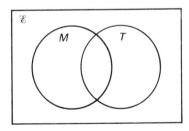

Complete it where:

$\mathscr{E}$ = {1, 2, 3, 4, 5, 6, 7, 8, 9, 10, 11, 12, 13, 14, 15, 16}

M = {multiples of 3}

T = {triangular numbers}

(b) Which multiples of 3 are triangular numbers?

(c) Which numbers are neither triangular numbers nor multiples of 3?

(d) Which numbers are not triangular numbers but are multiples of 3?

3

Exercise 4

Answer the following questions using the given Venn diagrams which show the numbers of members in a set:

1. $\mathscr{E}$ = {pupils in a certain form}
   C = {pupils who can play chess}
   S = {pupils who can play Scrabble}
   Find the number of pupils who:
   (a) can play chess,
   (b) can play Scrabble but not chess,
   (c) can play both chess and Scrabble,
   (d) can neither play Scrabble nor chess,
   (e) cannot play chess,
   (f) are in the form.

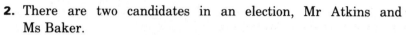

2. There are two candidates in an election, Mr Atkins and Ms Baker.
   $\mathscr{E}$ = the set of people who may vote
   A = the set of people who voted for Mr Atkins
   B = the set of people who voted for Ms Baker

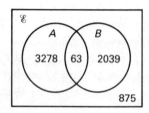

   If anyone voted for both candidates that vote was spoilt.
   (a) How many spoilt votes were there?
   (b) How many voted for Ms Baker only?
   (c) How many altogether voted for Mr Atkins?
   (d) How many did not vote?
   (e) How many voted?

Exercise 5

Draw Venn diagrams to show each of the following, then answer the given questions:

1. If the universal set $\mathscr{E}$ = {7, 8, 9, 10, 11, 12, 13, 14, 15, 16}, list the members of each of the following sets:
   (a) multiples of 5,   (b) prime numbers.

2. If $\mathscr{E}$ = {whole numbers less than or equal to 12}
   S = {square numbers} and
   M = {multiples of 3}, find:

   (a) the number of square numbers,
   (b) the number of members in set M,
   (c) the number of members that are not multiples of 3,
   (d) the numbers that are both square numbers and multiples of 3,
   (e) the numbers that are not square numbers,
   (f) the multiples of 3 that are not square numbers,
   (g) the square numbers that are not in set M,
   (h) the numbers that are in set S or set M or both.

3. $\mathscr{E}$ = {whole numbers that are less than 20}
   F = {2, 3, 4, 5, 6} G = {2, 4, 6, 8}
   Write whether the following are true or false:

   (a) G has 4 members.
   (b) The universal set $\mathscr{E}$ has 19 members.
   (c) 5 is in both F and G.
   (d) 8 is in G but not in F.
   (e) There are 16 numbers not in G.
   (f) There are 14 numbers not in F.
   (g) The number of elements that are in F or G or both totals 9.

4. $\mathscr{E}$ = {whole numbers that are less than 15}
   A = the set of even numbers B = {1, 2, 3, 4, 5, 6}
   Find:

   (a) the numbers that are in both A and B,
   (b) the numbers that are not in set A,
   (c) the numbers that are in A or B or in both,
   (d) the numbers that are not in A or in B.

5. $\mathscr{E}$ = {natural numbers less than 18}
   P = {prime numbers} D = {odd numbers}
   Find:

   (a) the numbers that are not in P,
   (b) the numbers that are in P or D or both,
   (c) the numbers that are in D but not in P.

Exercise 6

In this exercise, A and B are subsets* of the universal set $\mathscr{E}$.

For each question, illustrate your answer with a Venn diagram.

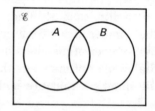

1. Is it true that the total number of elements in set A or B or both is always greater than the number of elements in both A and B?

2. Can the number of elements in both A and B be greater than the total number of elements in A or B or both?

3. Is it true that the number of elements in set A is always greater than the number of elements in both A and B?

4. Is it always true that the total number of elements in A or B or both is greater than the number of elements in A?

5. Is it always true that the number of elements that are not in set A is greater than the number of elements in set B?

Exercise 7   Logic Problems

1. In the Venn diagram,
   $\mathscr{E} = \{\text{pupils in Form 4G}\}$,
   $B = \{\text{pupils who play badminton}\}$,
   $S = \{\text{pupils who play squash}\}$.

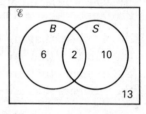

   Find:
   (a) the number of pupils who play both badminton and squash,
   (b) the number of pupils who play squash,
   (c) the number of pupils who play neither game,
   (d) the number of pupils who only play one of the games,
   (e) the number of pupils in the class.

*See the glossary, p.480.

2. The police had 39 suspects, all male. The guilty person was a fat man who was less than 1.75 m in height. Of the suspects, a total of 16 were fat, 15 were under 1.75 m in height and 8 were under 1.75 m in height but were not fat. Draw a Venn diagram where
   $\mathcal{E}$ = the set of suspects,
   F = {fat men} and
   H = {men under 1.75 m in height}.

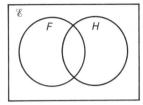

   Find the number of suspects who could have committed the crime.

3. Some pupils were asked whether they listened to the radio or watched TV the night before. Altogether, a total of 23 watched TV, 3 listened to the radio but did not watch TV, 8 did both and 9 did not watch TV.
   Show the information on a Venn diagram then find:
   (a) the number who did not listen to the radio,
   (b) the number who watched TV but did not listen to the radio,
   (c) the number who neither watched TV nor listened to the radio.

4. Last week, out of the 4th-year pupils, 84 went to the school dance, 29 went to the youth club as well as the dance, 74 pupils did not attend the dance while 101 did not go to the youth club. Draw a Venn diagram, then find:
   (a) the number that did not go to the dance or the youth club,
   (b) the total number who attended the youth club,
   (c) the number who went to the dance but did not go to the youth club.

5. In a survey, some people were asked if they took milk or sugar in their coffee. A total of 41 people said they took milk, a total of 45 took either milk or sugar or both, 24 took milk but no sugar while 5 took neither.
   Show the information on a Venn diagram.
   (a) How many took sugar?
   (b) How many took both milk and sugar?
   (c) How many did not take milk?

6. Mr and Mrs Allen arranged a party for their son. They bought some red and green balloons and gave all of them out. 5 children got both a red and a green balloon, 8 children got a red balloon but not a green one, 6 did not get a red balloon while 10 children did not get a green balloon.

Draw a Venn diagram then find:

(a) the total number who got a red balloon,

(b) the total number who got a green balloon,

(c) the number who did not get a balloon,

(d) the total number of children at the party,

(e) the total number of balloons bought by Mr and Mrs Allen.

7. A shop carried out a survey on the sort of fish bought by its customers. They discovered that 14% bought both cod and haddock, 26% bought haddock but not cod and 38% did not buy cod.

Show the above on a Venn diagram. Find:

(a) the total percentage who bought cod,

(b) the percentage who bought haddock,

(c) the percentage who bought neither cod nor haddock.

8. Some people were asked whether they would eat cabbage or sprouts. 91% said they would eat cabbage or sprouts or both, 25% would eat sprouts but not cabbage while 22% would not eat sprouts.

Draw a Venn diagram.

(a) What percentage would eat both cabbage and sprouts?

(b) What percentage altogether would eat sprouts?

(c) What percentage would eat cabbage but not sprouts?

2 Approximations, Estimations and Error

1. A bag of potatoes weighs 25.3 kg. Write this correct to the nearest kilogram.

2. Mrs Connelly's shopping came to £26.74. Write the cost correct to the nearest pound.

3. A piece of rope is 5.82 m long. Give its length correct to the nearest metre.

4. Give the following lengths correct to the nearest metre:
   (a) 8.4 m
   (b) 7.9 m
   (c) 2.65 m
   (d) 9.72 m
   (e) 14.49 m
   (f) 70.54 mm
   (g) 5.2 m
   (h) 560 cm
   (i) 825 cm
   (j) 196 cm
   (k) 3460 mm
   (l) 4752 mm

5. Write the following lengths correct to the nearest centimetre:
   (a) 5.6 cm
   (b) 8.1 cm
   (c) 42.7 cm
   (d) 6.44 cm
   (e) 19 mm
   (f) 563 mm

6. Write each mass correct to the nearest kilogram:
   (a) 6.7 kg
   (b) 2.3 kg
   (c) 1.84 kg
   (d) 21.2 kg
   (e) 52.65 kg
   (f) 9300 g
   (g) 4960 g
   (h) 7845 g

7. Write each mass correct to the nearest 10 g:
   (a) 73 g
   (b) 28 g
   (c) 91 g
   (d) 387 g
   (e) 512 g
   (f) 236 g
   (g) 4176 g
   (h) 6052 g
   (i) 96 g
   (j) 365 g
   (k) 1845 g
   (l) 6999 g

8. Write each mass correct to the nearest 100 g:
   (a) 487 g
   (b) 691 g
   (c) 449 g
   (d) 2699 g
   (e) 1248 g
   (f) 7060 g

9. Give the following correct to the nearest litre:

(a) 1.6 ℓ   (c) 55.5 ℓ   (e) 208.56 ℓ
(b) 8.35 ℓ   (d) 468.7 ℓ   (f) 99.93 ℓ

Exercise 2

1. Round to two decimal places:

(a) 4.867   (c) 1.293   (e) 0.6851   (g) 0.4165
(b) 9.177   (d) 0.306   (f) 12.166   (h) 2.0051

2. Round to one decimal place:

(a) 6.38   (c) 84.71   (e) 7.02   (g) 2.375
(b) 9.06   (d) 59.84   (f) 31.666   (h) 8.99

3. Round to two significant figures:

(a) 16.74   (c) 1.073   (e) 782   (g) 0.0499
(b) 2.86   (d) 0.782   (f) 0.499   (h) 299.6

4. Round to three significant figures:

(a) 1.648   (c) 1648   (e) 199.9   (g) 0.016 85
(b) 16.48   (d) 293.4   (f) 0.4772   (h) 0.054 96

5. A calculator gives the value of π as 3.141 592 7. Write this number correct to:

(a) 3 s.f.   (c) 4 s.f.   (e) 4 d.p.
(b) 2 d.p.   (d) 1 s.f.   (f) 2 s.f.

Exercise 3

Work these out mentally:

e.g. $150 \times 12 = \underline{1800}$ $\div 2$

(One method is: $150 \times 12 = 300 \times 6 = 1800$.)

$\times 2$

1. 47×10 **7.** 1000×8.2 **13.** 6×250
2. 73×10 **8.** 100×1.74 **14.** 125×8
3. 100×61 **9.** 25.6×100 **15.** 38×5
4. 10×546 **10.** 35×6 **16.** 75×16
5. 2.4×10 **11.** 4.5×12 **17.** 36×25
6. 7.9×100 **12.** 1.5×18 **18.** 2.25×24

Error

Suppose we wish to measure the length of a line segment AB using a ruler calibrated in units of 1 cm (that is, the divisions are 1 cm apart as shown on the first ruler in the diagram).

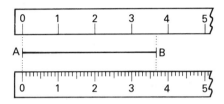

The length of AB lies between 3 cm and 4 cm. Since it is nearer 4 cm we say AB = 4 cm correct to the nearest centimetre.

If the actual length of AB is 3.68 cm then there is an *error** of 0.32 cm (4 − 3.68 = 0.32). 'Error' does not mean a mistake in this instance.

> The error in measuring is the difference between the recorded length and the actual length.

The second ruler shown in the diagram is calibrated in millimetres. More accurate measurements can be made since the divisions are smaller. It can be seen that AB lies between 3.6 cm and 3.7 cm but is nearer 3.7 cm, so we say AB = 3.7 cm correct to the nearest 0.1 cm (or 3.7 cm correct to the nearest mm). There is still an error but the error is much smaller. The error in this case is 0.02 cm (3.7 − 3.68 = 0.02).

Note When the unit of measurement is smaller, the error is usually smaller so a more accurate answer is obtained. Note also that a measurement of 5 cm suggests a measurement to the nearest centimetre while 5.2 cm is more accurate since it is a measurement to the nearest millimetre (or 0.1 cm). For the same reason, 5.0 cm is more accurate than 5 cm. This shows that, for a particular measurement, the greater the number of significant figures the more accurate the answer (since 5.0 cm has two significant figures whereas 5 cm has only one).

*See the glossary p.478.

1. Which is the more accurate measurement:
(a) 6 m or 6.3 m?   (f) 12.7 ℓ or 3.58 ℓ?
(b) 4.62 m or 4.6 m?   (g) 2.06 ft or 1.9 ft?
(c) 5.8 cm or 9 cm?   (h) 5.6 yd or 2.70 yd?
(d) 3 kg or 4.7 kg?   (i) 1.936 km or 8.74 km?
(e) 1.64 kg or 3.4 kg?   (j) 8.02 km or 8.020 km?

2. Calculate the error if:
(a) The actual length = 4.82 m and the recorded length = 4.8 m.
(b) The actual length = 12.6 cm and the recorded length = 13 cm.
(c) The actual length = 7.89 km and the recorded length = 8 km.
(d) The actual mass = 2.235 kg and the recorded mass = 2.2 kg.
(e) The actual mass = 5.608 kg and the recorded mass = 5.6 kg.
(f) The actual mass = 8.46 kg and the recorded mass = 8.5 kg

3. Calculate the error if:
(a) 3.64 km is rounded to 4 km
(b) 3.64 km is rounded to 3.6 km
(c) 4.87 km is rounded to 4.9 km
(d) 4.87 km is rounded to 5 km
(e) 6.28 km is rounded to 6 km
(f) 6.28 km is rounded to 6.3 km
(g) 9.31 km is rounded to 9 km

4. Calculate the error if:
(a) 5.43 is rounded to 1 s.f.
(b) 5.43 is rounded to 2 s.f.
(c) 6.152 is rounded to 1 s.f.
(d) 6.152 is rounded to 2 s.f.
(e) 6.152 is rounded to 3 s.f.
(f) 2.963 is rounded to 1 s.f.
(g) 2.963 is rounded to 2 s.f.
(h) 2.963 is rounded to 3 s.f.
(i) 71.39 is rounded to 3 s.f.
(j) 71.39 is rounded to 2 s.f.
(k) 71.39 is rounded to 1 s.f.

In estimating the answer to a calculation an error occurs.

The error in estimating is the difference between the estimation and the actual answer.

Exercise 5

Copy and complete the table:

	Actual calculation and answer	Estimated calculation and answer	Error
e.g. 1	49 × 31 = 1519	50 × 30 = 1500	⁻19*
e.g. 2	1.8 × 42 = 75.6	2 × 40 = 80	⁺4.4*
1.	8.1 × 5.7 =	8 × 6 =	
2.	24 × 66 =	20 × 70 =	
3.	7.9 × 13 =	8 × 10 =	
4.	93 × 760 =	90 × 800 =	
5.	47 ÷ 9.8 =	50 ÷ 10 =	
6.	588 × 83 =	600 × 80 =	
7.	62 ÷ 2.2 =	60 ÷ 2 =	
8.	390 ÷ 37 =	400 ÷ 40 =	
9.	196 ÷ 8.1 =	200 ÷ 8 =	
10.	65 × 7.4 =	70 × 7 =	
11.	210 ÷ 4.96 =	200 ÷ 5 =	
12.	578 × 92 =	600 × 90 =	

*Note An error of ⁻19 means that the estimated answer is 19 smaller than the actual answer while an error of ⁺4.4 means that the estimation is 4.4 larger than the actual answer.

Exercise 6

For any calculation, rounding *correct to one significant figure* usually gives a reasonable estimate—but does it always give the best estimate? Investigate this.

e.g. $4.8 \times 2.2 = 10.56$

By rounding correct to 1 s.f., we get $5 \times 2 = 10$
Error $= 10 - 10.56 = $ ‾$\underline{0.56}$

By rounding both numbers up, we get $5 \times 3 = 15$
Error $= 15 - 10.56 = \underline{4.44}$

By rounding both numbers down, we get $4 \times 2 = 8$
Error $= 8 - 10.56 = $ ‾$\underline{2.56}$

By rounding 4.8 down and 2.2 up, we get $4 \times 3 = 12$
Error $= 12 - 10.56 = \underline{1.44}$

So in this instance, rounding correct to 1 s.f. produces the best estimate (i.e., the smallest error).

A **1.** Write two numbers of your own.

2. Multiply the two numbers together to find an exact answer.

3. As in the example above, find the error created by rounding before multiplying if:
(*a*) both numbers are rounded to one significant figure,
(*b*) both numbers are rounded up,
(*c*) both numbers are rounded down,
(*d*) the first number is rounded up and the second one down,
(*e*) the second number is rounded up and the first one down.
(Note that one of the parts (*b*), (*c*), (*d*) or (*e*) will be identical to part (*a*), so there should only be four different answers.)

4. Which method of rounding gave the best estimate?

5. Repeat questions 1 to 4 above but use other pairs of numbers.

B Repeat section A for division.

C Repeat section A for addition.

D Repeat section A for subtraction.

Note Sometimes it is not necessary to round all numbers to one significant figure before estimating.

e.g. $3.45 \times 4.1 \approx 3.5 \times 4 = \underline{\underline{14}}$

In the example above, 3.45 has been rewritten as 3.5 for the estimation (two significant figures).

DO NOT WORK WITH MORE THAN ONE SIGNIFICANT FIGURE WHEN ESTIMATING UNLESS YOU ARE ABLE TO CARRY OUT THE ESTIMATION IN YOUR HEAD.

When calculating always check your answers by estimation.

Exercise 7

1. If 278 tickets were sold each day for 12 days, estimate the total number of tickets sold.

2. Estimate the time taken to travel 239 miles at 42 m.p.h.

3. Estimate the number of weeks from 14 June to 14 September of the same year.

4. Estimate the number of weeks it would take to save £98 if £3.90 were saved each week.

5. Estimate the number of weeks it would take to save £136 if £6.85 were saved each week.

6. A family's gas bill works out to be about £125 a quarter (i.e., a quarter of a year—three months). Estimate how much that is per week.

7. If a carton of milk costs 24 p and a milkman delivers 3 cartons per day to a certain house, estimate the milk bill for the month.

8. A carpet costs £10.75 per square metre. Estimate the cost of 18 m².

9. A bus set off at 11.17 and arrived at its destination at 13.46. Estimate the time for the journey.

10. Some tape is 18 mm wide. Estimate the number of strips of tape needed to cover a width of 236 mm.

3 Calculations and Number Patterns

Answer the questions in this exercise without using a calculator:

A 1. Write in figures:
   (*a*) thirty-nine,
   (*b*) two hundred and seventy-eight,
   (*c*) five thousand one hundred and six,
   (*d*) seven thousand and sixty-one,
   (*e*) eight hundred and twenty thousand, five hundred,
   (*f*) two million, three hundred and two thousand, four hundred and forty.

2. Write in words:
   (*a*) 465   (*b*) 30 029   (*c*) 581 300   (*d*) 2 034 070

B Find:

1. $46 + 87$
2. $429 + 578$
3. $73 - 25$
4. $624 - 156$
5. $4016 - 349$
6. $1 + 4 + 7 + 10 + 13 + 16$
7. $518 + 79 + 673$
8. $7 + 1066 + 299$
9. $49 + 77 - 36$
10. $53 - 26 + 34$

11. 17×4
12. 3×56
13. 6×320
14. 41×32
15. 143×25
16. $150 \div 5$
17. $730 \div 2$
18. $3560 \div 40$
19. $559 \div 13$
20. $1904 \div 34$

C Find the missing number:

1. $46 + 27 = 34 + \boxed{?}$
2. $72 - 25 = 52 - \boxed{?}$
3. $61 - 38 = \boxed{?} - 40$
4. $21 + 44 = 24 + \boxed{?}$

5. $19 + 61 = \boxed{?} + 45$ **7.** $87 - 51 = 90 - \boxed{?}$

6. $55 - 29 = \boxed{?} - 19$ **8.** $112 - 76 = \boxed{?} - 100$

D 1. Find the sum of 96 and 58.

2. Calculate the difference between 106 and 67.

3. What is the product of 7 and 29?

4. Multiply 82 by 4.

5. Take 59 from 92.

6. A number, when divided by 4, gives an answer of 8. What is the number?

7. A number, when divided by 7, gives an answer of 19. What is the number?

8. A number, when divided by 6, gives an answer of 5 and a remainder of 3. Find the number.

E Find the value of:

1. $13 + (7 - 2)$ **6.** $(12 + 7) - (4 + 6)$
2. $21 - 9 - 4$ **7.** $3 \times (8 - 5)$
3. $21 - (9 - 4)$ **8.** $3 \times 8 - 5$
4. $7 + 2 \times 8$ **9.** $39 - 4 \times 7$
5. $(7 + 2) \times 8$ **10.** $14 + 3 \times 6$

Exercise 2

Here are the first thirteen square numbers written in order:

1   4   9   16   25   36   49   64   81   100   121   144   169

Now   Take the first from the third: $9 - 1 = 8$
Take the second from the fourth: $16 - 4 = \boxed{?}$
Take the third from the fifth: $25 - 9 = \boxed{?}$
Take the fourth from the sixth and so on until you take the eleventh from the thirteenth.

Look carefully at your list of answers.
What can you say about these numbers?

17

Exercise 3

$$63 \times 48 = 3024$$
$$\text{and} \quad 36 \times 84 = 3024$$

The same answer is obtained.

Note that 63 was changed to 36
and 48 was changed to 84.

Check that 69×32 gives the same answer as 96×23.

Use a calculator to help you to find some more 2-digit numbers where this happens (i.e., by reversing the digits and then multiplying, the same answer is obtained).

Try to explain how to recognise that two 2-digit numbers give the same answer when reversed and multiplied.

Exercise 4

Do not use a calculator in this exercise.

A Which is bigger:

1. 4.6   or 3.9?
2. 52.8 or 7.965?
3. 2.899 or 2.988?
4. 7.0992   or 7.099 09?
5. 0.064   or 0.083?
6. 0.007 63 or 0.012?

B Find:

1. 4.2 + 0.94
2. 0.6 + 0.04
3. 0.525 + 0.747
4. 4.69 + 0.7
5. 2.83 + 16.2 + 0.45
6. 0.9 + 47.3 + 1.08
7. 23.81 − 8.65
8. 4.5 − 0.83
9. 3.51 × 4
10. 0.6 × 9
11. 2.8 × 7
12. 2.8 ÷ 7
13. 2.8 × 0.7
14. 2.8 ÷ 0.7
15. 3.7 × 0.4
16. 2.18 × 0.5
17. 1.913 × 0.7
18. 2.86 × 0.9
19. 4.14 × 0.3
20. 4.14 ÷ 0.3
21. 6.15 ÷ 0.5
22. 12.72 ÷ 0.4
23. 7.371 ÷ 0.9
24. 3.206 ÷ 0.7
25. 0.35 ÷ 0.2
26. 0.486 ÷ 0.6

27. 1.4×4.3

28. 36.7×0.21

29. $4.732 \div 1.3$

30. $0.3605 \div 0.35$

C **1.** Find the difference between 26 and 14.2.

2. Find the product of 0.8 and 0.7.

3. Find the sum of 2.104 and 0.987.

4. Multiply 70.2 by 5.

5. Take 2.07 from 10.91.

6. Add 1.9 to the product of 0.4 and 3.

D If $365 \times 14 = 5110$, find the value of:

1. 3.65×1.4

2. 36.5×140

3. 365×0.14

4. 3.65×0.14

5. 0.365×0.014

6. $5110 \div 365$

7. $511 \div 3.65$

8. $51.1 \div 1.4$

Approximations to Obtain Reasonable Answers

If 8 pencils cost £1.31 then 1 pencil costs £1.31 ÷ 8, so each pencil costs 16.375 p. This is a silly answer! Rounding to one decimal place, each pencil costs 16.4 p. This answer is still not sensible.

16 p is a reasonable answer for the cost of a pencil as 16.375 p gives 16 p when rounded to the nearest penny

Exercise 5

A Write each answer correct to two decimal places:

1. $7.53 \div 5$

2. $12.46 \div 8$

3. $53 \div 3$

4. $26 \div 7$

5. $2.6 \div 7$

6. $8.79 \div 4$

7. $27 \div 16$

8. $2.7 \div 16$

9. $0.27 \div 16$

10. $17 \div 6$

11. $17 \div 0.6$

12. $57.8 \div 1.1$

13. $4.71 \div 5.3$

14. $0.26 \div 0.29$

15. $0.7 \div 4.8$

B Write each answer correct to three significant figures:

1. $46 \div 9$	**6.** $7.4 \div 0.6$	**11.** $24.5 \div 1.9$
2. $10 \div 7$	**7.** $0.29 \div 0.8$	**12.** $76.3 \div 8.9$
3. $15 \div 8$	**8.** $1.23 \div 0.7$	**13.** $0.99 \div 0.41$
4. $8.6 \div 6$	**9.** $43 \div 1.2$	**14.** $0.58 \div 0.071$
5. $0.71 \div 4$	**10.** $179 \div 2.3$	**15.** $0.044 \div 0.57$

Exercise 6

A **1.** A running track is 9.9 m wide. If it is divided into 8 lanes of the same width, find the width of each lane correct to two decimal places.

2. A swimming bath is 10 m wide. If it is divided into 6 lanes of the same width, find the width of each lane correct to two decimal places.

3. A car travels 265 miles on 7 gal of petrol. Find the number of miles per gallon correct to one decimal place.

4. A cyclist travels 71 km in 4 h. Calculate the speed correct to two significant figures.

5. A rectangular room is 4.92 m long and 3.86 m wide. Calculate its area correct to two significant figures.

6. If Mrs Quinn pays out £7.95 per month, how much is the total payment for one year correct to the nearest pound?

7. If the bus fare for a journey of 18 km is 97 p, find the following correct to two decimal places:
(*a*) the cost, in pence, of travelling 1 km,
(*b*) the number of kilometres travelled for 1 p.

8. A parallelogram of area 2.54 m^2 has a base of length 1.7 m. Calculate its perpendicular height correct to one decimal place.

B Where appropriate, round an answer to make it sensible:

1. If 6 pencils cost 88 p, what does 1 pencil cost?

2. If 6 pencils cost £1.04, what does 1 pencil cost?

3. A running track is 8 m wide. It is divided into 7 lanes. How wide is each lane?

4. 12 sweets of equal mass weigh 125 g. Find the mass of 1 sweet.

5. A rectangle is 4.7 cm long and 3 cm wide. Calculate its area.

Exercise 7   Calculator Problems

The questions in this exercise *must* be answered using a calculator:

1. How many times can 62 be subtracted from 5394?

2. How many times can 124 be subtracted from 11 584?

3. What is the remainder when 3922 is divided by 47?

4. What is the remainder when 4161 is divided by 29?

5. Given that 43 is a factor of 2537, find another factor of 2537.

6. Given that 71 is a factor of 2627, find another factor of 2627.

7. Two 2-digit numbers when multiplied together give 3417. Find both numbers.

8. Two 2-digit numbers when multiplied together give 4753. Find both numbers.

9. After entering a number in my calculator, I multiplied by 3.2, added 4.6 and multiplied the answer by 3. I then subtracted 16 and divided the answer obtained by 4. If 10.25 was now in the display, find the number I started with.

10. After entering a number in my calculator, I divided by 3.4 then subtracted 1.8. I multiplied the answer by 8.2 then added 19.6. I then divided this answer by 1.8 and obtained 32.3 in the display. Find the number I started with.

Exercise 8   Decimal and Vulgar Fractions

A Answer these. Where possible, simplify your answers.

1. $1\frac{3}{4} + 2\frac{3}{4}$

2. $1\frac{1}{2} + 2\frac{3}{4}$

3. $4\frac{1}{3} + 2\frac{1}{6}$

4. $2\dfrac{3}{5} + 3\dfrac{7}{10}$

5. $3\dfrac{1}{4} + 1\dfrac{5}{6}$

6. $3\dfrac{5}{8} - 1\dfrac{1}{4}$

7. $4\dfrac{7}{8} - 1\dfrac{1}{2}$

8. $4\dfrac{2}{3} - 2\dfrac{5}{6}$

9. $3\dfrac{1}{3} - 1\dfrac{3}{5}$

10. $6\dfrac{3}{10} - 2\dfrac{3}{4}$

11. $6 \times \dfrac{3}{8}$

12. $\dfrac{3}{10} \times \dfrac{5}{6}$

13. $\dfrac{2}{3} \times 2\dfrac{1}{4}$

14. $2\dfrac{1}{10} \times 4\dfrac{1}{6}$

15. $3\dfrac{3}{8} \times 1\dfrac{7}{9}$

16. $2\dfrac{1}{2} \div 5$

17. $8 \div \dfrac{2}{3}$

18. $\dfrac{3}{4} \div \dfrac{5}{8}$

19. $6 \div 2\dfrac{2}{3}$

20. $4\dfrac{9}{10} \div 2\dfrac{4}{5}$

B 1. Find $\frac{1}{3}$ of 57.

2. The distance to school is $3\frac{1}{2}$ miles. After walking $1\frac{3}{8}$ miles of the journey to school, how much further must I walk?

3. I travelled $\frac{3}{4}$ of a journey of 84 km. How far was that?

4. Ruth spent $\frac{5}{6}$ of £54. How much was that?

5. Jayesh spent $\frac{3}{8}$ of his money. If he had £40 left,
(*a*) how much did he spend?
(*b*) how much did he have at first?

6. Sharon used $3\frac{5}{8}$ yards of material. If she then bought another $2\frac{1}{2}$ yards, how many yards was that altogether?

Exercise 9

A Change to decimals:

1. $\dfrac{7}{10}$

2. $\dfrac{4}{5}$

3. $\dfrac{1}{4}$

4. $\dfrac{7}{20}$

5. $\dfrac{17}{20}$

6. $\dfrac{3}{8}$

7. $\dfrac{6}{25}$

8. $\dfrac{9}{16}$

9. $\dfrac{23}{40}$

10. $\dfrac{16}{25}$

B Change to vulgar (common) fractions in their lowest terms:

1. 0.3   4. 0.14   7. 0.875   10. 0.068
2. 0.6   5. 0.88   8. 0.175   11. 0.004
3. 0.75   6. 0.54   9. 0.425   12. 0.8275

C Change to decimals giving each answer correct to three decimal places:

1. $\dfrac{1}{3}$ 3. $\dfrac{5}{6}$ 5. $\dfrac{5}{9}$ 7. $\dfrac{2}{7}$ 9. $\dfrac{11}{12}$

2. $\dfrac{2}{3}$ 4. $\dfrac{4}{9}$ 6. $\dfrac{8}{9}$ 8. $\dfrac{4}{7}$ 10. $\dfrac{11}{15}$

Exercise 10

1. Look at this method for finding a fraction that lies between two other fractions (not necessarily half-way between):

   e.g. 1 Find a fraction that lies between $\frac{1}{2}$ and $\frac{2}{3}$.

   $$\text{One possible fraction} = \frac{1+2}{2+3} = \underline{\underline{\frac{3}{5}}}$$

   (by adding the numerators and adding the denominators)

   e.g. 2 Find a fraction that lies between $\frac{5}{8}$ and $\frac{3}{4}$.

   $$\text{One possible fraction} = \frac{5+3}{8+4} = \frac{8}{12} = \underline{\underline{\frac{2}{3}}}$$

   (a) Check that $\dfrac{1}{2} < \dfrac{3}{5} < \dfrac{2}{3}$ (from e.g. 1)

   (b) Check that $\dfrac{5}{8} < \dfrac{2}{3} < \dfrac{3}{4}$ (from e.g. 2)

   (c) Try some fractions of your own.

   Does the method always work or can you find two fractions for which it doesn't work?

2. Using the method of question 1, which of the following is true for the new fraction?

   A It is always closer to the smaller of the two given fractions.

   B It is always closer to the larger of the two given fractions.

   C It is sometimes closer to the smaller fraction and sometimes closer to the larger fraction.

23

Directed Numbers

Exercise 11

A The average height of 12 people is 174 cm to the nearest centimetre. For each question find the difference between the given height and the average of 174 cm.

e.g. 1   Height = 185 cm   difference = $^+$11 cm

e.g. 2   Height = 171 cm   difference = $^-$3 cm

(*Note* $^-$3 cm means the height is 3 cm less than the average height.)

1. 168 cm	**3.** 169 cm	**5.** 180 cm	**7.** 186 cm	**9.** 160 cm
2. 182 cm	**4.** 183 cm	**6.** 173 cm	**8.** 178 cm	**10.** 153 cm

B Show the following as directed numbers and simplify each result:

e.g.   A lift goes up 3 floors then down 5 floors.

$^+3 + {}^-5 = {}^-2$

This is the same as going down 2 floors.

1. A lift goes up 2 floors then down 6 floors.

2. A lift goes down 1 floor then up 4 floors.

3. Anna gained 7 points but lost 11 points.

4. Ivan lost 8 points and gained 10 points.

5. Victor lost 9 points and gained 4 points.

6. I travel 32 km due north then 25 km due south.

7. I travel 45 km due north then 56 km due south.

8. I travel 56 km due south then 45 km due north.

9. Mrs Parr deposited £36 in the bank then later withdrew £51.

10. Mr Ross withdrew £62 from the bank then later deposited £29.

11. I made a loss of £114 then a profit of £86.

12. I made a profit of £66 then a loss of £100.

13. I made a profit of £87 then a loss of £58.

14. I made a loss of £92 then a profit of £143.

15. The aeroplane gained 650 m in height then descended 940 m.

16. The aeroplane descended 425 m then descended a further 390 m.

17. 36 people got off the train and 7 got on. At the next station, 49 got off and 12 got on.

18. 19 people got off the bus at one stop. A further 13 people got off at the next stop. No one got on at those stops.

19. A car uses 27 ℓ of petrol. It then uses a further 9 ℓ. 48 ℓ are put into the tank and a further 33 ℓ are used.

20. Mr O'Shea puts £53 into his bank account. He then withdraws £26 followed by another £45.

Exercise 12

Throughout this exercise let deposits be positive and withdrawals negative. For each question, find the overall deposit or withdrawal.

e.g. 1 Mrs Barrett withdrew £17 per month for 5 months.

Since $5 \times {}^-17 = {}^-85$, the overall withdrawal = £85.

e.g. 2 Mr Wilkinson deposited £26 per month for 4 months and withdrew £19 per month for 8 months.

Since $4 \times 26 + 8 \times {}^-19 = 104 + {}^-152 = 104 - 152 = {}^-48$, the overall withdrawal = £48

1. Mr Ashdown withdrew £23 per month for 3 months.

2. Mrs Brodie deposited £35 per month for 4 months.

3. Mrs Curley withdrew £56 per month for 5 months.

4. Donald withdrew £7 per month for 9 months.

5. Emily deposited £13 per month for 10 months.

6. Mr Furr withdrew £78 per month for 7 months.

7. Geraldine deposited £9 per month for 5 months then withdrew £6 per month for 7 months.

8. Henry deposited £8 per month for 6 months then withdrew £7 per month for 9 months.

9. Mr Ingham withdrew £14 per month for 3 months and deposited £20 per month for 4 months.

10. Mrs Jones withdrew £25 per month for 5 months and deposited £21 per month for 3 months.

11. Mr Kumar withdrew £19 per month for 3 months then withdrew a further £34 per month for 2 months.

12. Mrs Lee withdrew £61 per month for 6 months then withdrew a further £48 per month for 5 months.

Exercise 13

Find the value of:

1. (a) 7×13 (b) $7 \times {}^-13$ (c) ${}^-7 \times 13$ (d) ${}^-7 \times {}^-13$

2. (a) $\dfrac{24}{6}$ (b) $\dfrac{24}{{}^-6}$ (c) $\dfrac{{}^-24}{6}$ (d) $\dfrac{{}^-24}{{}^-6}$

3. (a) $2 \times 6 + 8$ (b) $2 \times {}^-6 + 8$ (c) ${}^-2 \times 6 + 8$

4. (a) $3 \times 9 - 4$ (b) $3 \times {}^-9 - 4$ (c) ${}^-3 \times {}^-9 - 4$

5. (a) $2(9 - 3)$ (b) ${}^-2(9 - 3)$ (c) $2(3 - 9)$

6. (a) $7 + 4 \times 5$ (b) $7 + 4 \times {}^-5$ (c) $7 + {}^-4 \times {}^-5$

7. (a) $2 \times 8 + 3 \times 7$ (b) $2 \times 8 + 3 \times {}^-7$
   (c) $2 \times {}^-8 + 3 \times {}^-7$

8. (a) $5 \times 6 + 4 \times 8$ (b) $5 \times {}^-6 + 4 \times 8$
   (c) ${}^-5 \times {}^-6 + 4 \times {}^-8$

26

Exercise 14   Sequences

Copy and complete these sequences by filling in the two missing terms in each:

1. 11, 15, 19, ? , 27, ?

2. 0.2, 0.4, 0.6, ? , ?

3. $1\frac{1}{2}$, 2, $2\frac{1}{2}$, ? , $3\frac{1}{2}$, 4, ?

4. 41, 38, ? , 32, ? , 26

5. $^-2$, $^-8$, $^-14$, $^-20$, ? , ?

6. 2, 5, 9, 14, ? , 27, ?

7. $8\frac{1}{4}$, 8, $7\frac{3}{4}$, $7\frac{1}{2}$, ? , ?

8. $^-1$, $^-3$, $^-6$, $^-10$, ? , ?

9. $^-2$, 4, $^-8$, 16, $^-32$, ? , ?

10. $^-1\frac{1}{2}$, $^-2\frac{1}{4}$, $^-3$, ? , $^-4\frac{1}{2}$, ?

Rules for Generating a Pattern or Sequence

Exercise 15

Give the first six terms of each sequence:

1. *Rule*   Double then add 5.
   Use the numbers {1, 2, 3, 4, 5, 6}
   Note   The sequence starts 7, 9, . . .

2. *Rule*   Double then take 1.
   Use the numbers {1, 2, 3, 4, 5, 6}

3. *Rule*   Multiply by 3 then subtract 4.
   Use the numbers {3, 4, 5, 6, 7, 8}

4. *Rule*   Multiply by 4 then add 3.
   Use the numbers {5, 6, 7, 8, 9, 10}

5. *Rule* Multiply by 6 then subtract 5.
  Use the numbers {3, 4, 5, 6, 7, 8}

6. *Rule* Multiply by 5 then add 9.
  Use the numbers {1, 2, 3, 4, 5, 6}

7. *Rule* Multiply by 3 then add 7.
  Use the numbers {4, 5, 6, 7, 8, 9}

8. *Rule* Multiply by 4 then take 6.
  Use the numbers {4, 5, 6, 7, 8, 9}

9. *Rule* Square then double.
  Use the numbers {1, 2, 3, 4, 5, 6}

10. *Rule* Square then multiply by 3.
  Use the numbers {3, 4, 5, 6, 7, 8}

The formula $T_n = 4n - 2$ also generates a sequence where T_n stands for the nth term of the sequence.
For example, the sixth term $T_6 = 4 \times 6 - 2 = 24 - 2 = 22$.

Exercise 16

Write the first six terms of each of the sequences given by the following nth terms:

1. $T_n = 4n - 2$ **5.** $T_n = 5n - 3$ **9.** $T_n = 6n - 5$

2. $T_n = 2n + 1$ **6.** $T_n = 5n + 2$ **10.** $T_n = 7n + 2$

3. $T_n = 3n + 3$ **7.** $T_n = 5n - 4$ **11.** $T_n = 4n^2$

4. $T_n = 5n - 1$ **8.** $T_n = 6n + 1$ **12.** $T_n = 6n^2$

Exercise 17

Try to find a rule or a formula for the nth term of each of the following sequences:

1. 4, 6, 8, 10, 12, 14, . . .

2. 9, 11, 13, 15, 17, 19, . . .

3. 3, 7, 11, 15, 19, 23, . . .

4. 5, 10, 15, 20, 25, 30, . . .

5. 7, 10, 13, 16, 19, 22, . . .

6. 0, 3, 6, 9, 12, 15, . . .

7. 6, 10, 14, 18, 22, . . .

8. 3, 8, 13, 18, 23, 28, . . .

Exercise 18   M

1. (*a*) Copy the table:

   (*b*) Complete your copy of the table by finding patterns in the rows and the columns.

a	b	c
4	1	13
6	3	27
8		41
	7	55
12	9	
	11	83
16		97
18	15	

2. (*a*) Copy the table:

   (*b*) Complete your copy of the table by finding patterns in the rows and the columns.

x	y	z
3	2	7
4		12
	6	19
6	8	28
7		
	12	
	14	67
10		

Exercise 19

Freddy the mathematical flea thought of a problem using part of a chess-board. The piece of chess-board was positioned with a corner at the top (see the following diagrams).

Freddy started at the top square and worked out the number of different paths to each of the other squares following two simple rules:

i. He only crossed the sides of the squares and not the vertices.

Crossing the vertices
is not allowed.

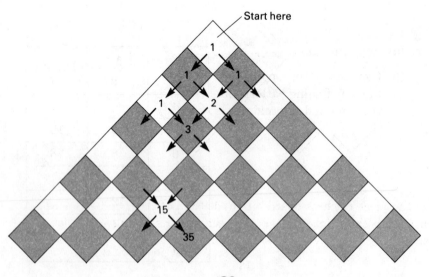

Start here

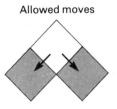

ii. He only moved downwards to a square below (from each square there are only 2 possible squares below to which he could move).

Note that there are 2 paths leading to the square labelled 2. These 2 paths are shown below:

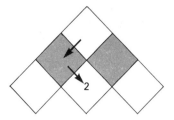 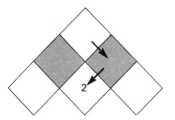

Note also that the square labelled 3 can only be reached by following the 3 paths shown.

Copy the piece of chess-board, at the bottom of p.30, and write in each square the total number of possible paths that lead to that square.

Blaise Pascal was born in Clermont-Ferrand, France, in 1623. His mother died when he was four and his father, fearing that his son would overtax his brain, insisted that he should only learn languages and that mathematics should be avoided.

Pascal began finding out about mathematics by himself and discovered many geometrical facts for himself including the fact that the sum of the angles of a triangle is equal to 2 right-angles. His father, on realising his son's ability, gave him a copy of Euclid's *Elements* (see Book 1G, p.279).

By the age of seventeen, Pascal had produced some original work on conic sections* which Descartes refused to believe had been written by a boy of that age. Pascal built the world's first calculating machine in 1642.

*See the glossary p.477

Pascal was once presented with a problem on gambling. He and another mathematician called Fermat worked on the problem. As a result of this, the ideas of probability were developed. Pascal used an 'arithmetic triangle' to help solve probability problems. The triangle has since been referred to as 'Pascal's triangle'. Questions on Pascal's triangle are given in Exercise 20.

On 23 November 1654 while Pascal was driving a carriage, the horses bolted and plunged from a bridge. Fortunately the harness snapped and the carriage stayed on the bridge. Pascal regarded his escaping death as a sign sent by God and from that moment on he devoted his life to religion.

Pascal suffered ill health throughout most of his life and at times was partly paralysed. Despite such problems he won fame as a mathematician and in the world of literature. He died in 1662 at the age of thirty-nine.

Exercise 20 ████████████████████████████ **M**

A Here is part of Pascal's triangle with some numbers missing. Copy it and try to work out and fill in the missing numbers (look for patterns):

Row 0						1					
Row 1					1		1				
Row 2				1		2		1			
Row 3			1		3		—		—		
Row 4		—		4		—		—		1	
Row 5	—		—		10		—		—		—
Row 6	1		—		—		20		—	6	—
Row 7	1	7		21		35		35	21	7	1
Row 8	—		—	28		—		—	56	—	—	—
Row 9	—	9	36		—		—		—	36	—	—
Row 10	—		—	45		—		—	252	210	—	—	—	—
Row 11	—		—	55		—	330		—	462	—	—	—	11	1

B Answer the following questions using Pascal's triangle:
 1. Find the sum of the numbers in each of the rows shown. For example, row 4 totals 16 since $1 + 4 + 6 + 4 + 1 = 16$.

2. (*a*) Ignoring the 1's at the end of each row, which rows contain only even numbers?

(*b*) If Pascal's triangle was continued beyond the rows given above, which would be the next row that would contain only even numbers (ignoring the 1's at the ends of the row)?

3. Note that all the numbers in row 5, except for the 1's at the ends, are divisible by 5.

(*a*) For the given rows of Pascal's triangle, write the row numbers where each number in the row, except for the 1's at the ends, is divisible by the row number. (There are 5 such rows including row 5.)

(*b*) What sort of numbers have you written for part (*a*)?

4. Find the values of 11^2, 11^3 and 11^4.
Compare the answers with Pascal's triangle.
Write what you notice.

C 1. In Pascal's triangle shown here, one sloping row has been marked.

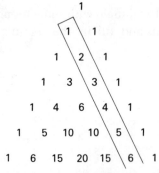

(*a*) Find the sum of the first two numbers in this sloping row and compare the answer with a number in the row below the second of the numbers.

(*b*) Find the sum of the first three numbers in this sloping row and compare the answer with a number in the row below the third of the numbers.

(*c*) Repeat the above for the first four numbers.

(*d*) Repeat the above for the first five numbers.

(*e*) Repeat the above for the first nine numbers in the larger Pascal's triangle obtained in part A.

(*f*) Write what you notice.

2. In this Pascal's triangle a different sloping row has been marked. What is the set of numbers in this row called?
(The name begins with the letter t.)

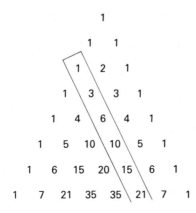

3. In this copy of Pascal's triangle a number of sloping rows have been marked.

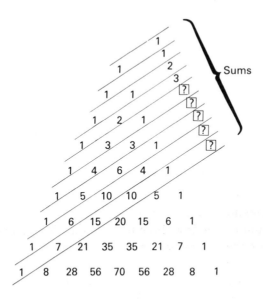

(a) Find the sum of the numbers in each sloping row (the first four have been added for you).
(b) The sums form a sequence. What is that sequence called?

35

D A Chinese version of Pascal's triangle is shown below. It is from a manuscript dated 1303, 320 years before the birth of Blaise Pascal.

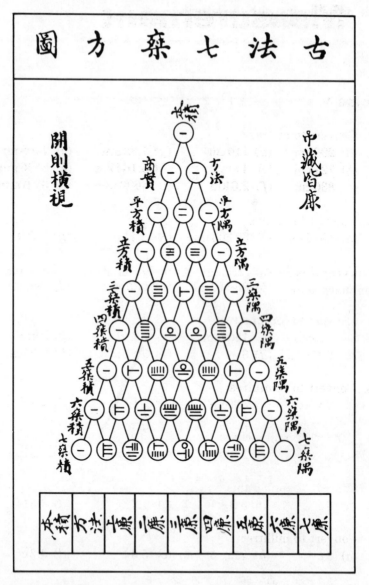

1. Use the triangle to find the symbols used for all the numbers from 1 to 10.

2. Write the symbols used for:

(a) 15　　(b) 20　　(c) 21　　(d) 28　　(e) 35　　(f) 56　　(g) 70

4 Units of Measurement

1. Convert to metres:
   - (a) 296 cm
   - (d) 110 cm
   - (g) 7.32 km
   - (j) 7409 mm
   - (b) 721 cm
   - (e) 4 km
   - (h) 5.164 km
   - (k) 5406 mm
   - (c) 830 cm
   - (f) 2.6 km
   - (i) 4600 mm
   - (l) 63 102 mm

2. Convert to grams:
   - (a) 4925 mg
   - (b) 37 200 mg
   - (c) 3.4 kg
   - (d) 9.72 kg

3. Convert to litres:
   - (a) 6400 mℓ
   - (b) 36 410 mℓ
   - (c) 835 cℓ
   - (d) 1980 cℓ

4. Convert to centimetres:
   - (a) 41 mm
   - (c) 150 mm
   - (e) 7.1 m
   - (g) 9 km
   - (b) 297 mm
   - (d) 6 m
   - (f) 3.82 m
   - (h) 2.5 km

5. Convert to centilitres:
   - (a) 24 mℓ
   - (b) 320 mℓ
   - (c) 6 ℓ
   - (d) 1.7 ℓ

6. Convert to millimetres:
   - (a) 6.3 cm
   - (b) 1.9 m
   - (c) 3.46 m
   - (d) 4.8 km

7. Convert to milligrams:
   - (a) 8 g
   - (b) 5.9 g
   - (c) 1.72 g
   - (d) 4.514 g

8. Convert to millilitres:
   - (a) 6 ℓ
   - (b) 21 ℓ
   - (c) 5.4 ℓ
   - (d) 3.2 cℓ

9. Convert to kilometres:
   - (a) 7000 m
   - (b) 6300 m
   - (c) 25 000 m
   - (d) 2810 m

10. Convert to kilograms:
    - (a) 6500 g
    - (b) 1941 g
    - (c) 3720 g
    - (d) 14 700 g

11. Convert to tonnes:

(*a*) 5000 kg   (*b*) 2800 kg   (*c*) 8160 kg   (*d*) 3045 kg

12. Convert to kilograms:

(*a*) 4 t   (*b*) 5.3 t   (*c*) 2.69 t   (*d*) 1.812 t

Exercise 2

1. Convert to metres:

(*a*) 2 km	(*d*) 0.436 km	(*g*) 2158 cm	(*j*) 844 mm
(*b*) 0.7 km	(*e*) 326 cm	(*h*) 9 cm	(*k*) 53 mm
(*c*) 0.19 km	(*f*) 49 cm	(*i*) 3512 mm	(*l*) 8 mm

2. Convert to grams:

(*a*) 823 mg   (*b*) 46 mg   (*c*) 0.8 kg   (*d*) 0.165 kg

3. Convert to litres:

(*a*) 741 mℓ   (*b*) 9 mℓ   (*c*) 36 cℓ   (*d*) 5 cℓ

4. Convert to centimetres:

(*a*) 6 mm   (*b*) 0.4 m   (*c*) 0.16 m   (*d*) 0.07 km

5. Convert to centilitres:

(*a*) 405 mℓ   (*b*) 0.23 ℓ   (*c*) 0.9 ℓ   (*d*) 0.06 ℓ

6. Convert to millimetres:

(*a*) 0.4 cm   (*b*) 0.2 m   (*c*) 0.67 m   (*d*) 0.07 m

7. Convert to milligrams:

(*a*) 6 g   (*b*) 0.6 g   (*c*) 0.05 g   (*d*) 0.019 g

8. Convert to millilitres:

(*a*) 0.1 ℓ   (*b*) 0.83 ℓ   (*c*) 0.06 ℓ   (*d*) 0.8 cℓ

9. Convert to kilometres:

(*a*) 9000 m   (*b*) 900 m   (*c*) 48 m   (*d*) 295 m

10. Convert to kilograms:

(*a*) 400 g   (*b*) 250 g   (*c*) 70 g   (*d*) 7 g

11. Convert to tonnes:

(*a*) 6000 kg   (*b*) 400 kg   (*c*) 36 800 kg   (*d*) 980 kg

12. Convert to kilograms:

    (*a*) 41.8 t   (*b*) 0.2 t   (*c*) 0.75 t   (*d*) 0.05 t

Addition and Subtraction

Exercise 3

1. A cyclist travelled the following distances: 4.6 km, 1.95 km, 830 m and 946 m. Find, in kilometres, the total distance travelled.

2. A piece of string that is 4.2 m long has the following lengths cut from it: 49 cm, 596 mm, 78 cm, 320 mm and 1.24 m. What length, in millimetres, remains?

3. A family used the following quantities of milk: 260 mℓ, 450 mℓ, 1.3 ℓ, 75 cℓ and 370 mℓ. How many litres was that altogether?

4. A baby had a mass of 3.18 kg at birth. At the age of 13 weeks the baby had a mass of 5.2 kg. Calculate the gain in mass.

5. An electrician used the following lengths of electric cable: 6.39 m, 9.28 m, 7 m, 8.4 m, 4.2 m and 5.08 m, calculate the total length used.

6. Mr Stevens bought 2 kg of sugar, 250 g of butter, 500 g of tomatoes, 2.5 kg of potatoes, 0.8 kg of apples and 120 g of mushrooms. Find the total mass of the shopping.

7. Sara used 230 mℓ from a 2 ℓ bottle of lemonade. How much was left?

8. 4.9 m of material is cut from a roll that is 36.8 m long. What length remains?

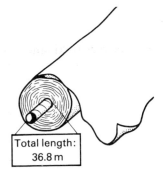

Total length: 36.8 m

Reminder

12 in (inches) = 1 ft (foot)	16 oz (ounces) = 1 lb (pound)
3 ft (feet) = 1 yd (yard)	14 lb (pounds) = 1 st (stone)
	8 pt (pints) = 1 gal (gallon)

Exercise 4

1. Chris needs two pieces of cord, one 2 ft 5 in long and the other 4 ft 2 in. Give the total length needed in feet and inches.

2. Tracy needs two pieces of cord, one 3 ft 8 in long and the other 4 ft 6 in long. Give the total length needed in feet and inches.

3. Julian weighs 9 st 5 lb, Robert weighs 8 st 12 lb while Melissa weighs 6 st 9 lb.
   (*a*) How much heavier than Melissa is Robert?
   (*b*) How much heavier than Melissa is Julian?
   (*c*) Find their total mass.

4. 2 gal 5 pt of oil was used from a drum containing 5 gal. How much oil was left in the drum?
   (*a*) Give your answer in gallons and pints.
   (*b*) Give the same answer in pints.

5. A piece of material of length 4 ft is cut from a roll that is 7 yd long. Find the length of the piece that is left, giving your answer:
   (*a*) in feet,   (*b*) in yards and feet.

Multiplication and Division

Exercise 5

1. Mr Elliot needed 6 pieces of wood, all 1.28 m long. Find the total length needed.

2. A box of sweets has a mass of 237 g. Find, in kilograms, the mass of:
   (*a*) 10 boxes   (*b*) 8 boxes   (*c*) 25 boxes

3. Angie pours 215 mℓ of orange into each of 7 glasses. How many litres of orange did she use?

40

4. 6 crème eggs have a total mass of 234 g. Find:

  (*a*) the mass of 1 egg,

  (*b*) the mass of 12 eggs,

  (*c*) the mass of 30 eggs, giving your answer in kilograms.

5. A piece of rope is 29.25 m long.

  (*a*) Find the length of each piece if the rope is cut into 5 equal lengths.

  (*b*) How many 4-metre lengths can be cut from the rope and what length would be left over?

6. (*a*) How many pieces of string each 28 cm long can be cut from a ball of string of length 30 m?

  (*b*) If the string is cut as in part (*a*), what length, in millimetres, is left over?

7. Which is heavier, 4 boxes each weighing 3.89 kg or 3 boxes each weighing 5.19 kg?

8. A bottle of bubble bath holds 330 mℓ. If a capful is used for each bath and if a capful holds 15 mℓ, how many baths can be had from one bottle?

Exercise 6

1. I used 9 pieces of tape, each of length 7 in. Find the total length in feet and inches.

2. How long would it take to fill a 5 gal drum at the following rates?

  (*a*) 4 pt per minute,

  (*b*) $\frac{1}{4}$ pt per minute.

3. Find the mass of 10 boxes of sweets, in pounds and ounces, if each box weighs:

  (*a*) 8 oz (*b*) 10 oz.

4. A piece of wood is 3 ft 4 in long.

  (*a*) How many 5-inch lengths can be cut from it?

  (*b*) If it was cut into 7-inch lengths, how many pieces would be obtained and what length, in inches, would be left over?

5. 15 pieces of rope are needed, each of length 4 ft. Find the total length needed in yards.

Miscellaneous Questions

Exercise 7

1. How many centimetres are there in 4.95 m?

2. Write 4.92 kg in grams.

3. Change 2470 mℓ to litres.

4. Which of the following is 350 mm written in metres?
   A. 35 000 m   B. 35 m   C. 3.5 m   D. 0.35 m   E. 0.035 m

5. Which is the heaviest?
   A. 60 g   B. 0.06 kg   C. 0.60 g   D. 0.6 kg   E. 6000 mg

6. Each side of a regular hexagon is 6.5 cm long. Which of the following is the length of its perimeter?
   A. 52 cm   B. 52 mm   C. 39 cm   D. 39 mm   E. 6.5 m

7. Angle iron is put on all 12 edges of a wooden box as shown. Find the total length used if the box is 2.4 m long, 1.8 m wide and 0.9 m deep.

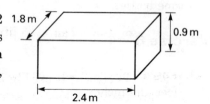

8. If a common brick weighs 2.58 kg find the mass of:
   (*a*) 10 bricks   (*b*) 100 bricks   (*c*) 500 bricks   (*d*) 800 bricks

9. A common brick weighs 2.58 kg and a hod (used for carrying bricks) has a mass of 5.45 kg. Find the total mass carried by a labourer if he carries:
   (*a*) 6 bricks in his hod,
   (*b*) 10 bricks in his hod.

10. A bus company wants to make 4 bus stops so that the buses stop in a straight line, one behind the other. What total length is needed if each bus is 12 m long and if a 3 m gap is allowed between each bus?

Exercise 8

1. A milk bottle holds 600 mℓ of milk. What is the volume of the bottle in cubic centimetres?

2. What is the volume, in cubic centimetres, of a bottle that holds 2 ℓ?

3. A four-wheel lorry has an unladen mass of 5 t. If it carries a load weighing 6.84 t, what is its laden mass?

4. An aeroplane carries 5 crew and 108 passengers together with their luggage. If the passengers' luggage has an average mass of 18 kg and if the people on the aircraft have an average mass of 67 kg, calculate the total mass carried. (Assume the crew do not have any luggage.) Give the answer in tonnes.

5. Mr and Mrs Grundy wish to make a shelf to stand 3 pans on in a straight line. If the diameters of the pans are 16 cm, 18 cm and 20 cm and if a 1.5-centimetre gap is allowed between each pan as well as a 2-centimetre space at each end, how long must the shelf be?

6. How many times can a 15-litre tank be filled from a 135-litre drum?

7. If 250 kg of cement is used in laying 1000 bricks:
   (a) How many bricks can be laid using 2 t of cement?
   (b) How many bricks can be laid per kilogram of cement?

8. A lorry carrying 50 bags of cement weighs 11.5 t, but when the same lorry carries 80 bags of cement it weighs 13 t. Find:
   (a) the mass of one bag of cement if they are all the same size,
   (b) the mass of the empty lorry.

5 Tables, Clocks and Dials

Australia

CHARGE BAND 5B

Time difference between 8 and 10 hours later than GMT

International Code 010	Country Code 61	Area Code	Subscriber's Number

Adelaide	8	Kalgoorlie	90
Albury	60	Launceston	03
Ballarat	53	Melbourne	3
Bathurst	63	Mt. Gambier	87
Brisbane	7	Newcastle	49
Bunbury	97	Orange	63
Bundaberg	71	Perth (W.A.)	9
Cairns	70	Rockhampton	79
Canberra	62	Sydney	2
Darwin	89	Toowoomba	76
Dubbo	68	Townsville	77
Geelong	52	Wagga Wagga	69
Geraldton	99	Wollongong	42
Hobart	02		

The table above shows the dialling codes for places in Australia. The code for Newcastle, Australia, is 010 61 49. Write the code for:

1. Wagga Wagga,
2. Kalgoorlie,
3. Darwin,
4. Canberra,
5. Wollongong,
6. Launceston,
7. Melbourne,
8. Sydney.

Exercise 2

The table gives the dimensions of two different chairs:

SC56 upright armchair		
SC49 upright chair		
Comfortable steel-framed chairs with upholstered seats and backs		
	mm	in
Seat height	450	$17\frac{3}{4}$
Overall height	810	32
SC49 overall width	490	$19\frac{1}{4}$
SC56 overall width	560	22
Overall depth	530	21

1. What is the overall height of the chairs in:
   (a) millimetres?
   (b) inches?

2. What is the seat height of the chairs in:
   (a) millimetres?
   (b) inches?

3. What is the difference between the overall height and the seat height in:
   (a) millimetres?
   (b) inches?

4. Which model (give its reference number) is the wider and by how many millimetres?

5. (a) The overall depth of 530 mm and 21 in suggests that:

$$21\,\text{in} = 530\,\text{mm}$$

$$\therefore \quad 1\,\text{in} = \frac{530}{21}\,\text{mm}$$

Carry out the calculation $\frac{530}{21}$ to find the number of millimetres in 1 inch (using these entries from the table).

(*b*) Carry out a similar calculation (as in part (*a*)) using the overall width of the SC56.

(*c*) Carry out similar calculations using the other pairs of figures in the table to find, for each pair, the number of millimetres in 1 inch.

(*d*) Try to explain why the answers for parts (*a*) to (*c*) give different numbers of millimetres in 1 inch.

Exercise 3   M

The table shows that number of pupils absent each day during four weeks of a school term:

Pupils absent from school

	week 1	week 2	week 3	week 4
Mon	58	61	69	71
Tues	52	60	74	72
Wed	54	64		82
Thur	62	66	73	
Fri	64	66	67	79
Totals			358	
Average		63.4		77.2

1. Copy and complete the table.

2. Use the table to find:
(*a*) the number absent on Thursday of week 4,
(*b*) the number absent on Wednesday of week 3.

3. If 843 pupils attended the school shown in the table of absences, draw another table, but this time, show the number of pupils present each day.

Timetables and Travel

Exercise 4

The timetables below give the train times between Holyhead and Euston:

	B	A		A			A		A			A	
Holyhead	—	0537	0600	0600	0705	—	1010	—	1246	1320d	—	1620	1803
Bangor	—	0606	0629	0629	0732	0921b	1038	1250	—	1417	—	1650	1831
Llandudno Junction	—	0624	0647	0647	0756	0940f	1059	1310	—	1438	1601c	1711	1850
Chester	0625	0730	0802	0802	0905	1056	1205	1403	1416	1552	1722	1819	2000
Crewe	0659	0804	0845	0845	0939	1130	1241	1438	1440	1626	1756	1854	2034
Watford Junction	—	—	1031	—	—	—	—	—	—	1803	1951	2048	2229
London Euston	1029	1010	1052	1132	1211	1400	1457	1710	1710	1829	2012	2109	2250

		A			A			G	A				A	B
London Euston	0650	0745	0855	1000	—	1100	1145	1300	1400	1430	1545	—	1703	1740
Watford Junction	0706	0801	0911	1016	1041	—	1201	—	—	—	—	1701	—	—
Crewe	0905	0938	1104	1215	1304	1350	1520	1601	1702	1802	1905	1905	2058
Chester	0939	1012	1138	1249	1342	1342	1425	1556	1635	1739	1837	1941	1941	2137
Llandudno Junction	1059f	1117	1250	—	1446	1446	1529	1707	1738	1847	1949	2101	2101	—
Bangor	—	1138	1315	1353	1507	1507	1549	1737g	1759	1907	2010	2121	2121	—
Holyhead	—	1210	—	1430	1539	1539	1620	1819g	1831	1938	2042	2153	2153	—

		A	
London Euston	1805	1900	2050
Watford Junction	—	—	—
Crewe	2024	2108	2312
Chester	2058	2142	2346
Llandudno Junction	2210	2301	0054c
Bangor	2238	2322	0114c
Holyhead	—	2354	0150c

Note Applicable throughout Watford and Euston Timetables
A Through service Holyhead to Euston or vice versa
B Through service Chester to Euston or vice versa
G Saturdays Only until 7th September
b Not Saturdays. Change at Chester
c Change at Chester

d Change at Bangor
f Change at Chester, except on Saturdays
g Change at Llandudno Junction

Use the timetable to help you to answer these questions:

1. If you catch the 07.05 train from Holyhead, at what time would you arrive at London Euston?

2. If you catch the 12.46 train from Holyhead, at what time would you arrive in Chester?

3. If you catch the 18.05 train from London Euston, at what time would you arrive in Chester?

4. At what time must you leave Holyhead to arrive at London Euston at 14.57?

47

5. What train must you catch from London Euston to get to Bangor at 15.49?

6. If you travel from Watford Junction and want to arrive at Llandudno Junction at 11.17, which train must you catch?

7. If you leave London Euston at 17.40 and arrive at your destination after 9 p.m., at which station will you get off the train?

8. If you travel on the 18.03 from Holyhead to Chester how long does the journey take?

9. If you catch the 15.56 from Chester to Bangor, at which station must you change trains?

10. Dan travelled to London Euston on a Saturday and had to change at Chester, from which station did he set off?

Exercise 5

The timetable below gives the times of trains travelling from Sheffield to Luton on a Saturday. Use it to answer the questions given.

Sheffield	Derby	Nottingham	Loughborough	Leicester	Kettering	Wellingborough	Luton
1051	1135	—	1154	1207	1232	1240	1308
—	—	1255	1309	1322	1345	1352	1452
—	1355	—	1414	1427	1452	1500	1552
1500	1541	—	—	1607	—	—	1656
—	—	1550	1604	1617	1643	1650	1752
1600	—	1657	—	1721	—	—	1810
—	1700	—	1719	1732	1757	1805	1857
—	—	1812	1826	1839	1905	1912	2012
1925	—	2022	—	2046	—	—	2135
2035	2116	—	2132	2145	2211	2218	2246
2359	0124	—	0143	0209	0305	0317	0355

1. If you leave Sheffield at 16.00, at what time would you arrive in Luton?

2. If you leave Nottingham at 15.50, at what time would you arrive in Kettering?

3. A train arrives in Wellingborough at 22.18, at what time does it leave Derby?

4. A train leaves Loughborough at 01.43, how long does it take to get to Luton?

5. A train arrives in Luton at 20.12, how long did it take to travel from Leicester?

Exercise 6

The following table gives some car-hire rates in America:

Car class	Hourly rate($)	Daily rate($)	Weekly rate($)	Extra days($)	Monthly rate($)
H	N/A*	N/A*	99.00	29.95	276.00
A	10.50	31.95	109.00	31.95	356.00
B	11.25	33.95	139.00	33.95	476.00
C	12.00	35.95	149.00	35.95	556.00
D	13.25	39.95	159.00	39.95	596.00
E	14.00	41.95	169.00	41.95	636.00
F	15.00	44.95	229.00	44.95	796.00
G	14.00	41.95	189.00	41.95	676.00
J	14.25	42.95	209.00	42.95	716.00
Q	15.00	44.95	229.00	44.95	796.00
Y	14.00	41.95	169.00	41.95	636.00

Note   *H-class cars cannot be booked for a period less than 6 days. For any extra hours the hourly rate will be $10.50. (N/A = non-applicable)

Answer the following questions using the table above:

1. Find the cost of hiring a D-class car for 1 week.

2. Find the cost of hiring a G-class car for 1 month.

3. Which two classes of car are the most expensive to hire?

4. Find the cost of hiring a C-class car for 3 days.

5. For an E-class car, after how many hours does hiring at the hourly rate become more expensive than hiring for 1 full day at the daily rate?

6. How much more does it cost to hire a J-class car for 7 days at the daily rate than for 1 week at the weekly rate?

7. Find the cost of hiring an A-class car for 9 days (i.e., for 1 week plus 2 extra days).

8. Find the cost of hiring a Y-class car for 17 days (i.e. 2 full weeks and 3 extra days).

9. Using the exchange rate: £1 = $1.43, find the cost of hiring a B-class car for 1 week giving the cost in pounds sterling and rounding the answer to the nearest penny.

10. If £1 = $1.43, find the cost of hiring an H-class car for 1 month giving the cost in pounds sterling and rounding the answer to the nearest penny.

Exercise 7 M

A Draw the following scales. Choose your own sizes. Complete each one.

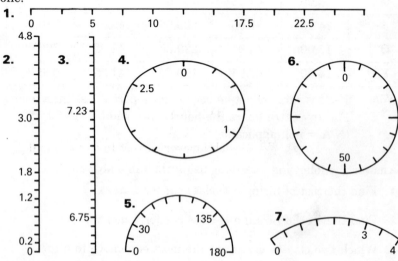

8.

9.

10.

B *On your copy* of the scales, draw pointers (or arrows) to show the following. (Draw the answer to question 1 on part A, question 1; the answer to question 2 on part A, question 2, etc.)

1. 15   **3.** 6.97   **5.** $67\frac{1}{2}$ **7.** 2.75   **9.** 3

2. 2.5   **4.** 2.25   **6.** $67\frac{1}{2}$ **8.** $102\frac{1}{2}$ **10.** 6.25

Exercise 8

Give the readings from the following scales, clocks and dials:

1. Gas cooker controls:

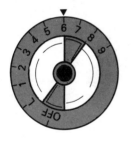

What gas mark is the cooker set at?

2. Camera-speed selector:

What speed is the camera set at, if 30 stands for $\frac{1}{30}$ s?

3. Measuring jug:

How many millilitres of water are in the jug?

4. Measuring jug:

How many pints of water are there in the jug?

51

5. Car speedometer:

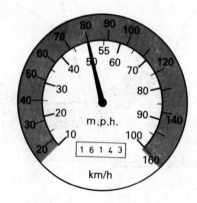

How fast is the car travelling?
(*a*) in m.p.h.?   (*b*) in km/h?

6. Tyre-pressure gauge:

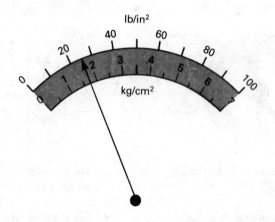

What is the tyre pressure in
(*a*) lb/in²?   (*b*) kg/cm²?

7. A clinical thermometer:

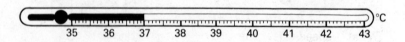

What temperature is shown?

8. An ammeter:

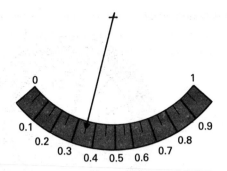

What is the reading in amps?

Exercise 9

For each dial, write the number the pointer has just passed. (Careful! Some pointers will have turned clockwise and some anticlockwise.)

1.

4.

7.

10.

2.

5.

8.

11.

3.

6.

9.

12.

Electricity meters show how much electricity has been used. The unit of electricity used is the kilowatt hour (kW h). Some meters have a digital display (they use figures) while others have dials. The diagrams below show both types:

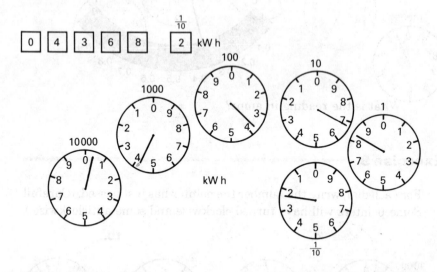

Both sets of meters show that 4368.2 units of electricity have been used. The $\frac{1}{10}$ kW h digit is usually ignored and so the readings would be given as 4368 kW h.

Exercise 10

Give the readings on the following meters in kilowatt hours (ignore the $\frac{1}{10}$ kW h digit):

1.

| 0 | 2 | 9 | 3 | 7 | $\frac{1}{10}$ 3 | kW h |

2.

| 0 | 7 | 0 | 2 | 6 | $\frac{1}{10}$ 5 | kW h |

3.

| 0 | 8 | 9 | 1 | 4 | $\frac{1}{10}$ 9 | kW h |

4.

| 1 | 4 | 5 | 9 | 2 | $\frac{1}{10}$ 8 | kW h |

5.

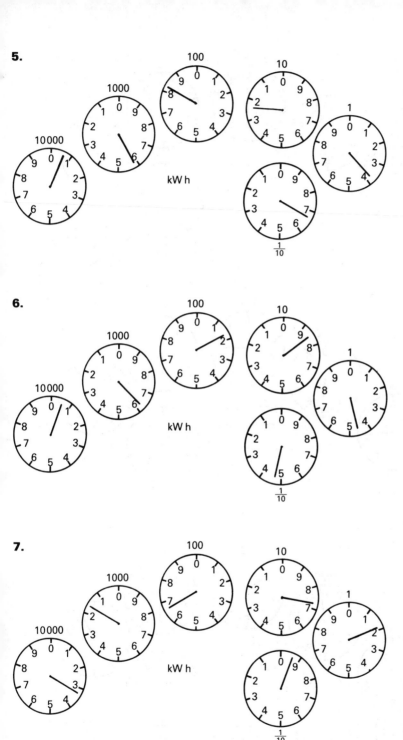

kW h

6.

kW h

7.

kW h

55

8.

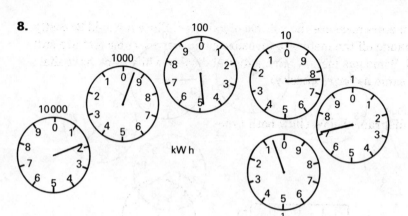

9.

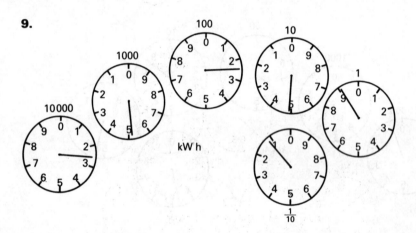

10.

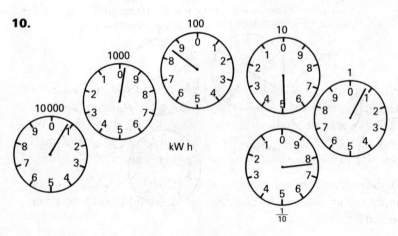

56

Gas meters measure the volume of gas used. Since it would be costly to change all the meters to measure cubic metres, cubic feet are still used. Some gas meters have a digital display while some have dials (the same as for electricity).

The diagrams below show both types.

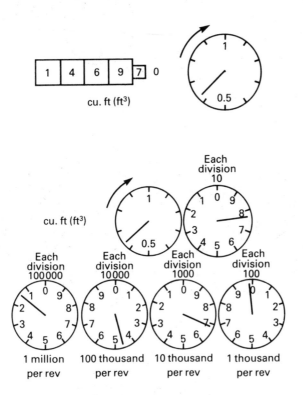

cu. ft (ft³)

The readings are given in hundreds of cubic feet of gas. So on the digital display, only the first four digits are read while on the dials, only the lower four dials are read (that is, the dials that give readings of 100 cubic feet or more). The pointer on the differently marked dial turns one revolution for every cubic foot of gas used.

Both meters above give a reading of 146 970 ft³. Normally the last two digits are ignored so the readings above would usually be given as 146 900 ft³.

Exercise 11

Give the readings on the following meters in hundreds of cubic feet of gas. (Ignore the tens and units digits.)

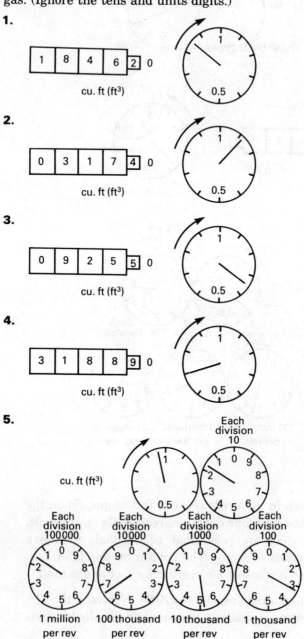

1.

| 1 | 8 | 4 | 6 | 2 | 0 |

cu. ft (ft³)

2.

| 0 | 3 | 1 | 7 | 4 | 0 |

cu. ft (ft³)

3.

| 0 | 9 | 2 | 5 | 5 | 0 |

cu. ft (ft³)

4.

| 3 | 1 | 8 | 8 | 9 | 0 |

cu. ft (ft³)

5.

cu. ft (ft³)

Each division 10

Each division 100000 — 1 million per rev

Each division 10000 — 100 thousand per rev

Each division 1000 — 10 thousand per rev

Each division 100 — 1 thousand per rev

6.

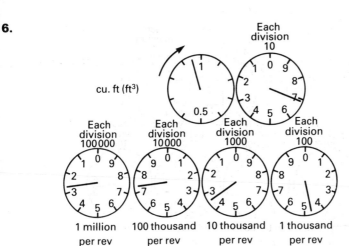

cu. ft (ft³)

Each division 10

Each division 100 000 — 1 million per rev

Each division 10 000 — 100 thousand per rev

Each division 1000 — 10 thousand per rev

Each division 100 — 1 thousand per rev

7.

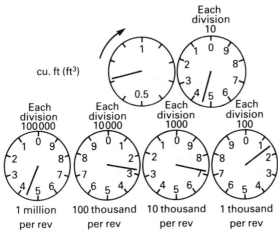

cu. ft (ft³)

Each division 10

Each division 100 000 — 1 million per rev

Each division 10 000 — 100 thousand per rev

Each division 1000 — 10 thousand per rev

Each division 100 — 1 thousand per rev

8.

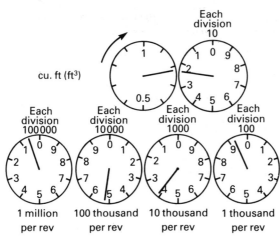

cu. ft (ft³)

Each division 10

Each division 100 000 — 1 million per rev

Each division 10 000 — 100 thousand per rev

Each division 1000 — 10 thousand per rev

Each division 100 — 1 thousand per rev

9.

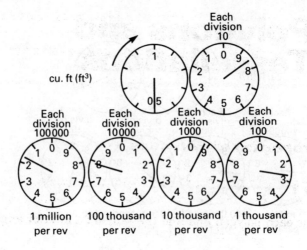

10.

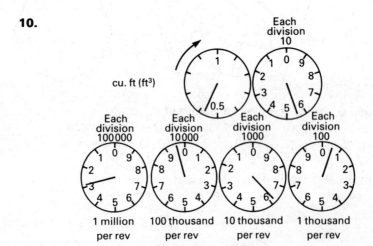

6 Polygons and Tessellations

Angle Properties of Polygons

Exercise 1

1. Name each polygon and find the sum of its interior angles:

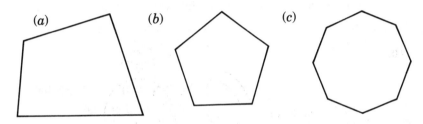

(a)   (b)   (c)

2. Find the sum of the interior angles of:
   (a) a hexagon,
   (b) a 7-sided polygon,
   (c) an 11-sided polygon,
   (d) a 16-sided polygon.

3. Find the missing angle:

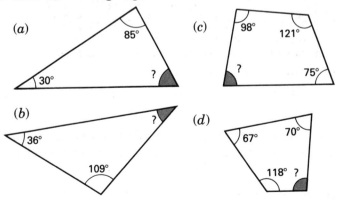

(a)   85°   30°   ?

(c)   98°   121°   ?   75°

(b)   36°   109°   ?

(d)   67°   70°   118°   ?

4. Two angles of a triangle measure 61° and 87°. Find the third.

5. Two angles of a triangle measure 114° and 39°. Find the third.

6. Three angles of a quadrilateral measure 84°, 142° and 59°. Calculate the fourth angle.

7. If three angles of a quadrilateral measure 98°, 67° and 88°, calculate the fourth angle.

8. If three angles of a quadrilateral measure 107°, 76° and 90°, calculate the fourth angle.

Exercise 2

Reminder: The sum of the exterior angles of a polygon = 360°.

A Calculate each interior angle of the following regular polygons:
  1. a pentagon,   **3.** a 15-sided polygon,
  2. a 12-sided polygon,   **4.** a 16-sided polygon.

B Calculate each exterior angle of the following regular polygons:
  1. a hexagon,   **3.** an 18-sided polygon,
  2. a 10-sided polygon,   **4.** a 30-sided polygon.

C Calculate the missing angle:

  1.   **2.**

   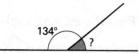

D Find the interior angles of regular polygons which have exterior angles measuring:
  1. 60°   **2.** 45°   **3.** 20°   **4.** 18°   **5.** 15°

E Check your answers to part A by first calculating each exterior angle then using that answer to help you to find each interior angle.

Exercise 3

Consider various regular polygons. For each one, write its number of sides and find the size of its exterior angles.

Now try to explain how to work out the number of sides of a regular polygon from the size of the exterior angles.

For a regular polygon:

$$\text{Number of sides} = \frac{360}{\text{size of exterior angle}}$$

Exercise 4

A Find, where possible, the number of sides of a regular polygon when its exterior angles measure:

1. 40°	**3.** 45°	**5.** 120°	**7.** 5°
2. 20°	**4.** 18°	**6.** 50°	**8.** 15°

B Find, where possible, the number of sides of a regular polygon when its interior angles measure:

1. 120°   **2.** 90°   **3.** 170°   **4.** 40°   **5.** 178°

Exercise 5

Carry out these constructions:

A *To Construct a Regular Pentagon:*

1 Draw a circle of any size.

2. Draw one radius.

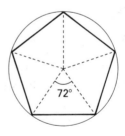

3. Since a regular pentagon would create 5 equal angles at the centre, each angle $= \dfrac{360°}{5} = 72°$.

Draw an angle of 72° to the radius from the centre of the circle so that the new arm of this angle cuts the circle.

4. Repeat step 3 on each new radius until 5 points are marked on the circle.

5. Join the 5 points to form a pentagon.

B Construct:

1. a regular octagon,
2. a regular decagon (a 10-sided polygon),
3. a regular 9-sided polygon.

Exercise 6

ABCDE is a regular pentagon centre O.

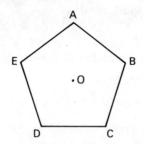

1. How many axes of bilateral symmetry has the pentagon?

2. What is its order of rotational symmetry?

3. If the pentagon is rotated clockwise about O so that vertex A moves to position B, what is the size of the angle of rotation?

4. If the pentagon is rotated clockwise about O so that vertex D moves to position B, what is the size of the angle of rotation?

5. If the pentagon is rotated anticlockwise about O through 144°,
   (a) into which lettered position will vertex E move?
   (b) into which lettered position will vertex A move?

6. (a) Calculate angle DOC.
   (b) If OD and OC are joined, what sort of triangle is DOC?
   (c) Calculate angle ODC.
   (d) Calculate angle EDC.
   (e) Calculate angle DEC.
   (f) Calculate obtuse angle BOD.
   (g) Calculate reflex angle COE.

Tessellations

Exercise 7

A Equilateral triangles tessellate.

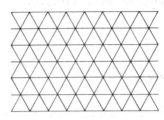

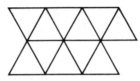

1. What is the size of each interior angle of an equilateral triangle?

2. What is the sum of the angles at a point?

3. How many equilateral triangles meet at a point?

4. Use your answers to the questions above to help you to explain why equilateral triangles tessellate.

B **1.** What is the size of each interior angle of a regular pentagon?

2. Will regular pentagons tessellate?

3. The diagram shows three regular pentagons meeting at a point. What is the size of the gap in degrees?

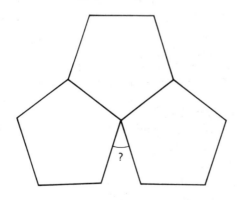

C Which regular polygons tessellate?

65

Each interior angle in a regular 9-sided polygon measures 140°, while each angle in an equilateral triangle measures 60°.

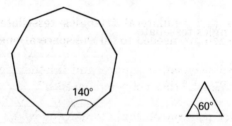

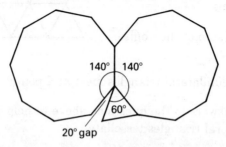

These two shapes do not tessellate together, since 140° + 140° + 60° = 340° and a 20° gap remains.

Neither will 9-sided polygons and squares tessellate together, since 140° + 140° + 90° = 370°. (They overlap.)

Exercise 8

All polygons referred to in this exercise are regular.

1. Which other polygon tessellates with octagons? (Two octagons and this other polygon fill the space at one point.)

2. Which polygon tessellates with 12-sided polygons? (Two 12-sided polygons and this other polygon fill the space at one point.)

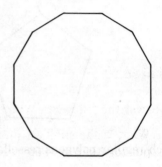

3. Squares and equilateral triangles can be used together to make a tessellation. How many of each are needed to fill the space at one point?

4. Hexagons and equilateral triangles tessellate together. How many of each are needed to fill the space at one point?

5. A 12-sided polygon, a square and another polygon tessellate together. Which other polygon is needed?

6. An 18-sided polygon, an equilateral triangle and another polygon can fill the space at one point. Which other polygon is it?

Polyominoes

This is a *monomino*:

Here is a *domino*:
(It is made up of two mono-minoes.)

There are only two types of *tromino*:
(Each tromino is made up of three monominoes.)

and

These are *not* trominoes (since at least one side of each square must touch a full side of a different square):

Note   There is only one possible monomino and one possible domino.

Exercise 9

A Answer the following. When an answer is possible draw a diagram.

1. What is the smallest number of straight trominoes that can be put together to make a square?

67

2. What is the smallest number of L-shaped trominoes that can be put together to make a square?

3. Show how to make a 4 × 4 square using:
  (*a*) 5 straight trominoes and 1 monomino.
  (*b*) 5 L-shaped trominoes and 1 monomino.

4. Show how to cover a 5 × 5 square using:
  (*a*) 8 straight trominoes and 1 monomino,
  (*b*) 8 L-shaped trominoes and 1 monomino.

5. Is it possible to cover a chess-board using the following?
  (Each square of the shape must cover a single square of the chess-board.)
  (*a*) only straight trominoes together with 1 monomino,
  (*b*) only L-shaped trominoes together with 1 monomino.

B Here is a *tetromino*:
(It is made up of four monominoes.)

1. Try to find all the possible tetrominoes.
   Draw them on dotty paper.

Note that these are said to be the same tetromino (if one is turned over, it will fit exactly on to the other):

2. What is the smallest possible square that can be made using only the T-shaped tetromino?

3. Try to make squares using the other tetrominoes. Use only one shape of tetromino at a time and in each case try to make the smallest possible square.

4. Try to make a 4 × 4 square using 4 of one type of tetromino. Do this five times using a different shaped tetromino each time. Draw your answers on dotty paper.

5. Try to cover a 5 × 5 square using 6 of one type of tetromino and 1 monomino. Do this five times using a different shaped tetromino each time. Draw your answers on dotty paper. (Note that the tetrominoes may be turned over before being used.)

C There are 12 *pentominoes* (each made up from five monominoes). Four pentominoes are shown here:

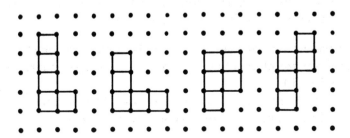

1. Try to find the other eight pentominoes (use dotty paper).

2. How many of the 12 pentominoes are nets of an open box? Show which are nets.

3. Using all 12 pentominoes, try to make:
(*a*) a 10 × 6 rectangle,   (*c*) a 15 × 4 rectangle,
(*b*) a 12 × 5 rectangle,   (*d*) a 20 × 3 rectangle.

4. Use a full set of pentominoes to make two separate rectangles. 6 pentominoes are used for each rectangle. The rectangles should be 6 squares long and 5 squares wide.

5. Use some of the pentominoes to make these rectangles. Decide which pentominoes to use each time.
(*a*)   5 by 3,   (*d*) 10 by 4,
(*b*)   5 by 4,   (*e*) 10 by 5.
(*c*) 10 by 2,

6. (*a*) Try to fit all 12 pentominoes on to a chess-board (having the same-sized squares as the pentominoes) such that each square of each pentomino covers a single square of the chess-board. (You may need to make a chess-board of a suitable size.)
(*b*) How many squares remain uncovered?

Exercise 10   Three Games

Game A (Use one set of pentominoes and a chess-board with squares of the same size as the squares of the pentominoes.)

  i. Two players share one set of pentominoes (6 each).

 ii. The players take turns to place a pentomino on the chess-board with squares on top of squares. (The pentominoes are not allowed to overlap.)

iii. You lose if you are the first player who cannot place a pentomino on the board without any overlapping.

Game B

  i. All 12 pentominoes are placed on the table.

 ii. Play continues in the same way as in game A except that each player may choose any of the pentominoes that have not been played on to the chess-board.

iii. You lose if you are the first player who cannot place a pentomino on the board without any overlapping.

Game C

  i. Follow Game A or B until play stops.

 ii. The player whose turn it is to move may move exactly one piece on the chess-board to a new position on the chess-board and then play a pentomino that has not yet been used.

iii. The other player now moves as explained in step ii.

iv. You lose if you are unable to place a pentomino on the board without overlapping.

Exercise 11

1. This pentomino tessellates:

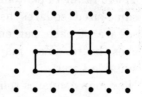

Here is the tessellation:

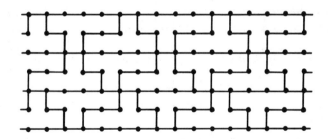

The same pentomino could have tessellated in another way to give a different pattern. Draw this different tessellation.

2. Does the L-pentomino tessellate?
If so, draw the tessellation.

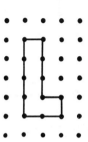

3. Does the P-pentomino tessellate?
If so, draw the tessellation.

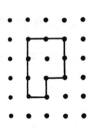

4. Test these other pentominoes to see if they tessellate.
If they do then draw the tessellation.

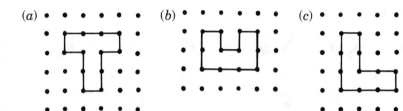

(a)   (b)   (c)

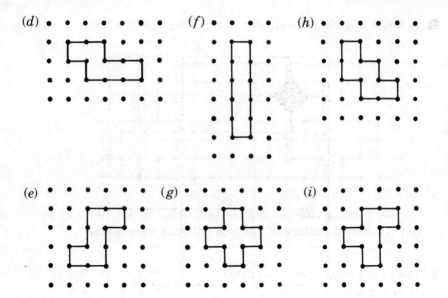

(d) (f) (h)

(e) (g) (i)

Exercise 12

A Challenge
Here is a hexomino:

There are 35 different hexo-
minoes.
Try to find them (use dotty paper
or squared paper).

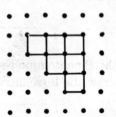

Exercise 13   Geometric Patterns in Two
Dimensions 　　　　　　　　　　　　　　　　　　 M

A Copy and complete the given pattern:

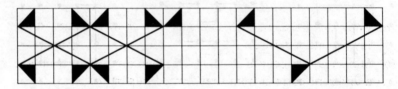

B 1. Here is a tile. The other tiles have the same design.

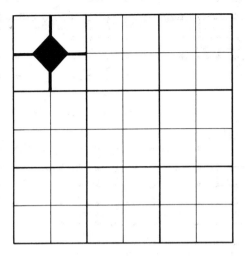

Copy the given diagram to show all the tiles.

2. Suppose you had 9 identical tiles. One is shown here.

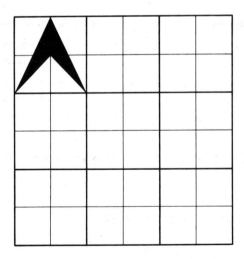

(a) Copy the diagram, then complete it to show how you would lay the other 8 tiles.

(b) Make another copy of the diagram, then complete it to obtain a different pattern from that formed in part (a).

Types of Number, Indices and Square Roots

7

Types of Number

When you were first learning to count you used the *natural numbers**.

The set of natural numbers = {1, 2, 3, 4, 5, . . .}.

You learnt to add and this simply used the natural numbers (adding two natural numbers gives another natural number).

Later, you learnt to 'take away'. You discovered, for example, that if you had 25 p and then spent 25 p you would have no money left since 25 − 25 = 0. When you had learnt the new symbol 0 (zero) you knew the *whole numbers*.

The set of whole numbers = {0, 1, 2, 3, 4, 5, . . .}.

Some subtraction questions give answers that are not whole numbers:

e.g. 3 − 7 = ⁻4

The answer is a negative number and this introduces a new set of numbers called *integers*.

The set of integers = {. . ., ⁻3, ⁻2, ⁻1, 0, 1, 2, 3, . . .}.

The set of integers can be shown on a number line where each number has a unique point:

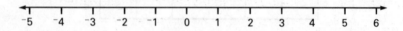

Multiplication can be carried out within the set of integers.

An integer × an integer = an integer

*See the glossary p.478.

74

However, division creates some new numbers called *rational numbers*:

e.g. 1 $\quad 8 \div 2 = 4$

e.g. 2 $\quad 2 \div 8 = \frac{1}{4}$ (or 0.25)

The second answer is not an integer.

A rational number is a number that can be written as a ratio of two numbers where the second of the two numbers is not zero.

$\frac{1}{4}$ is obviously a rational number.

4 can be written as $\frac{4}{1}$ and is therefore a rational number.

$2.75 = 2\frac{3}{4} = \frac{11}{4}$ and is therefore a rational number.

In fact: All natural numbers are whole numbers.
           All whole numbers are integers.
           All integers are rational numbers.

Rational numbers can be shown on a number line where each number has a unique point. Some are shown below:

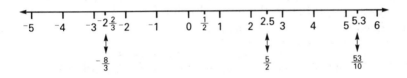

The set of rational numbers that lie between 0 and 1

$$= \{\tfrac{1}{2}, \tfrac{1}{3}, \tfrac{1}{4}, \tfrac{1}{5}, \tfrac{1}{6}, \tfrac{1}{7}, \tfrac{1}{8}, \ldots\}.$$

We can continue for ever listing members of this set.

Note that recurring decimals are rational numbers.

e.g. $0.142\,857\,142\,8\ldots = 0.\dot{1}42\,85\dot{7} = \frac{1}{7}$ which is rational.

When we find square roots we come across some numbers that are not rational. They are called *irrational numbers**. When written as decimals they continue for ever and do not recur. $\sqrt{2}, \sqrt{3}, \sqrt{5}, \sqrt{6}, \sqrt{7}, \sqrt{8}, \sqrt{10}$ are all irrational numbers. $\sqrt{4}$ and $\sqrt{9}$ are rational numbers since $\sqrt{4} = 2 = \frac{2}{1}$ and $\sqrt{9} = 3 = \frac{3}{1}$.
π is also an irrational number.

*See appendix 1 p.471.

75

Exercise 1

1. Find $\sqrt{2}$ on a calculator and note the answer.

2. Multiply the answer obtained in question 1 by itself (i.e., square it).

In question 2 of Exercise 1 you probably did not get the answer 2. The answer would probably be 2.000 000 1 or 1.999 999 8 or 1.999 999 9 which shows that the calculator value for $\sqrt{2}$ was not exact even though it was probably given to seven decimal places.

Consider the right-angled triangle ABC shown here:

$AC^2 = 1^2 + 1^2$

$\qquad = 1 + 1 = 2$

so $AC = \sqrt{2}$.

This shows that although the decimal value for $\sqrt{2}$ continues for ever a fixed length $\sqrt{2}$ units long can be drawn.

The diagram below shows that other square roots, most of them irrational numbers, can each be shown by a line of fixed length.

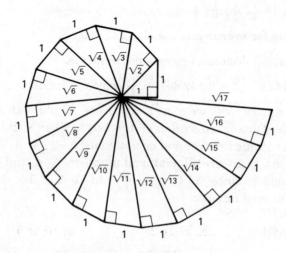

This number line shows some rational and irrational numbers:

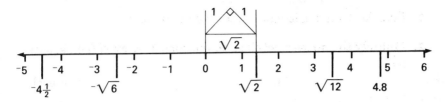

The irrational numbers $-\sqrt{6}$, $\sqrt{2}$ and $\sqrt{12}$ are shown on the line above.

Exercise 2

Here is a set of numbers:

$$\{6, \ ^-2, \ 0, \ 1.8, \ \sqrt{3}, \ ^-6.5, \ \tfrac{1}{3}, \ \sqrt{16}, \ 2\tfrac{1}{2} \ ^-\sqrt{5}, \ \sqrt{21}, \ 9, \ ^-2.2\}$$

From the set above, list the set of:

1. natural numbers,
2. integers,
3. rational numbers,
4. positive integers,

5. whole numbers,
6. negative integers,
7. non–negative integers,
8. irrational numbers.

Indices

Exercise 3

A Find the value of:

1. (a) 2^2 (b) 10^2 (c) 8^2 (d) 11^2 (e) 12^2

2. (a) 8^3 (b) 9^3 (c) 11^3 (d) 4^4 (e) 2^9

3. (a) $2^3 \times 3^2$ (b) $5^2 \times 3^3$ (c) $7^2 \times 2^4$

4. (a) $\dfrac{6^3}{2^2}$ (b) $\dfrac{5^4}{2^3}$ (c) $\dfrac{9^2}{10^3}$

B Which has the larger value?

1. 4^3 or 3^4 2. 8^2 or 2^8 3. 9^4 or 5^6

C Use a calculator to find the value of:

1. 6^4
2. 7^4
3. 14^2
4. 19^2
5. 15^3
6. 4^6
7. 3^8
8. 11^4
9. 14^5
10. $5^4 \times 2^3$
11. $6^3 \times 7^4$
12. $3^6 \times 8^3$

D Copy these, but write the correct sign ($<$, $>$ or $=$) in place of each question mark:

1. 4^5 ? 3^6
2. 2^7 ? 7^2
3. 6^2 ? 2^4
4. 3^7 ? 25^2
5. 2^{12} ? 13^3
6. 9^5 ? 3^{10}
7. 2^{18} ? 6^6
8. 20^3 ? 5^6
9. 18^4 ? 9^6

E Work these out on a calculator. Give each answer correct to four significant figures:

1. 1.87^2
2. 2.08^2
3. 0.92^2
4. 0.778^2
5. 68.1^2
6. 39.5^2
7. 971^2
8. 2.4^3
9. 1.42^3
10. 0.57^4
11. 1.26^4
12. 63^3
13. 1.01^6
14. 0.93^8
15. 4.49^5

Exercise 4

A Find the value of:

1. $(^-3)^2$
2. $(^-7)^2$
3. $(^-10)^2$
4. $(^-12)^2$
5. $(^-3.5)^2$

B Find the value of:

1. 2×9^2
2. 2×4^2
3. 3×2^2
4. $2 \times (^-4)^2$
5. $2 \times (^-5)^2$
6. 4×7^2
7. $3 \times (^-6)^2$
8. 5×8^2
9. $4 \times (^-8)^2$
10. 6×6^2
11. $3 \times (^-1)^2$
12. $5 \times (^-9)^2$

Exercise 5

1. If $y = x^2$, find the value of y when x equals:
   (a) 9
   (b) 0
   (c) $^-1$
   (d) $^-7$
   (e) 10

2. If $y = 2x^2$, find the value of y when x equals:
   (a) 3
   (b) 7
   (c) $^-1$
   (d) $^-6$
   (e) $^-2$

78

3. If $y = 4x^2$, find the value of y when x equals:

(a) 1   (b) $^-1$ (c) 0   (d) 5   (e) $^-5$

4. If $y = x^2 + 3$, find the value of y when x equals:

(a) 2   (b) $^-6$ (c) 0   (d) 5   (e) $^-5$

5. If $y = x^2 + 7$, find the value of y when x equals:

(a) 5   (b) $^-5$ (c) 3   (d) $^-3$ (e) $^-8$

6. If $y = x^2 - 4$, find the value of y when x equals:

(a) 0   (b) $^-6$ (c) 6   (d) 7   (e) $^-7$

Exercise 6

1. If $a = 5$, find the value of:

(a) $3a$ (b) a^3 (c) $a + a + a$ (d) $a \times a \times a$

2. If $p = 3$, find the value of:

(a) $4p + p$ (b) $4p^2$ (c) $5p$ (d) $5p^2$

3. If $m = 6$, find the value of:

(a) $5m + 2m$ (b) $5m \times 2m$ (c) $10m^2$ (d) $7m$

4. If $k = {}^-2$, find the value of:

(a) $2k + 4k$ (b) $6k$ (c) $6k^2$ (d) $2k \times 4k$

5. If $x = {}^-9$, find the value of:

(a) $x + 2x$ (b) $2x^2$ (c) $3x$ (d) $x \times 2x$

6. If $g = 4$, find the value of:

(a) $2g$ (b) $2g^2$ (c) $(2g)^2$ (d) $4g^2$

7. If $t = 7$, find the value of:

(a) $3t$ (b) $3t^2$ (c) $(3t)^2$ (d) $9t^2$

8. If $c = {}^-3$, find the value of:

(a) $2c$ (b) $2c^2$ (c) $(2c)^2$ (d) $4c^2$

Exercise 7

1. If $A = l^2$ gives the area of a square with side l cm, find the area of a square with side:

(a) 7 cm   (b) 4.5 cm   (c) 2.8 m

2. If $a = 4l^2$, find a when l equals:
  (*a*) 5   (*b*) 7   (*c*) 3.5

3. $V = l^3$. Find V when l equals:
  (*a*) 4   (*b*) 9   (*c*) 2.1

4. $A = 3r^2$ will give an approximate value of the area of a circle with radius r units. Find the area of a circle with radius:
  (*a*) 4 cm   (*b*) 6 cm   (*c*) 4.7 cm

5. $A = \frac{3}{4}d^2$ gives an approximate value of the area of a circle with diameter d units. Find the area of a circle with diameter:
  (*a*) 8 cm   (*b*) 9 cm   (*c*) 1.8 m

6. The formula $V = 4r^3$ gives an approximate value of the volume of a sphere. Find the volume of a sphere if its radius, r, equals:
  (*a*) 2 cm   (*b*) 5 m   (*c*) 3.2 m

Exercise 8

Simplify, leaving answers in index form:

A
1. $x^2 \times x^5$ **5.** $d^2 \times d^8$ **9.** $k^5 \times k^9$

2. $a^3 \times a^6$ **6.** $g^5 \times g$ **10.** $9^2 \times 9^6$

3. $m^7 \times m^2$ **7.** $p \times p^7$ **11.** $e^{11} \times e^5$

4. $4^8 \times 4^7$ **8.** $u^8 \times u^7$ **12.** $w^{14} \times w^{15}$

B
1. $t^3 \times t^4 \times t^2$ **5.** $b^2 \times b^4 \times b^4$ **9.** $h^7 \times h \times h^2$

2. $f^6 \times f^2 \times f^3$ **6.** $z \times z^6 \times z^2$ **10.** $q^9 \times q^5 \times q$

3. $c^9 \times c^2 \times c^4$ **7.** $v^8 \times v^4 \times v^2$ **11.** $y^{10} \times y^4 \times y^6$

4. $l^5 \times l^6 \times l^4$ **8.** $n^5 \times n^2 \times n^5$ **12.** $3^2 \times 3^2 \times 3^7$

C
1. $x^4 \times y^3 \times y^2 \times x^5$ **5.** $u^8 \times u^2 \times p^3 \times p$

2. $c^2 \times g^3 \times c^3 \times g^4$ **6.** $z^6 \times m \times m^5 \times m^4$

3. $n^6 \times v^2 \times v^4 \times n^3$ **7.** $b^5 \times b^2 \times f^4 \times b^2$

4. $k^2 \times q^4 \times q^3 \times k^4$ **8.** $l^6 \times l^7 \times h^9 \times l^3$

9. $w^3 \times x^3 \times x^2 \times w^3$ **11.** $d^2 \times d^2 \times p^5 \times t^4 \times t^3$

10. $a^8 \times m^4 \times m^3 \times n^4 \times a^2$ **12.** $u^5 \times d^3 \times u^5 \times c^2 \times u^5$

Exercise 9

A Simplify:

1. $2x \times 3x$ **5.** $5a^4 \times a^3$ **9.** $6g \times 3g$

2. $3p^2 \times 5p^3$ **6.** $3t^2 \times 2t$ **10.** $4l \times l^2$

3. $2v \times 4v^2$ **7.** $8c^2 \times c$ **11.** $2n^6 \times 5n^2$

4. $4m^3 \times 2m^2$ **8.** $3k^3 \times 4k$ **12.** $4q^4 \times 7q^7$

B **1.** $c^3d^2 \times c^2d^4$ **5.** $2g^2h^3 \times 3g^3h^2$

2. $p^2q \times p^2q^4$ **6.** $2nu^3 \times 4u^2n$

3. $e^4k^3 \times e^2k$ **7.** $5p^3c \times 3p^4c^2$

4. $m^2t^4 \times t^2m^3$ **8.** $4v^6w^2 \times v^3w^4$

Exercise 10

A Simplify the following, leaving your answers in index form:

1. $\dfrac{x^8}{x^2}$ **4.** $\dfrac{x^9}{x^6}$ **7.** $\dfrac{x^9}{x^2}$ **10.** $5^{10} \div 5^6$

2. $\dfrac{x^4}{x}$ **5.** $\dfrac{2^7}{2^4}$ **8.** $x^{10} \div x^4$ **11.** $\dfrac{7^6}{7^5}$

3. $\dfrac{x^5}{x^4}$ **6.** $\dfrac{3^8}{3^4}$ **9.** $x^3 \div x^2$ **12.** $x^8 \div x^5$

B Express each of the following as a single power, *then* find its value:

1. $\dfrac{2^5}{2^2}$ **4.** $\dfrac{6^7}{6^4}$ **7.** $\dfrac{8^9}{8^8}$ **10.** $\dfrac{3^3}{3^5}$

2. $\dfrac{4^7}{4^5}$ **5.** $\dfrac{4^9}{4^5}$ **8.** $\dfrac{2^{15}}{2^6}$ **11.** $\dfrac{4^8}{4^9}$

3. $\dfrac{9^6}{9^4}$ **6.** $\dfrac{3^8}{3^3}$ **9.** $\dfrac{2^5}{2^7}$ **12.** $\dfrac{10^6}{10^9}$

Exercise 11

A Simplify the following by expressing each one as a single power:

1. $\dfrac{x^5 \times x^7}{x^6}$

2. $\dfrac{a^9 \times a^3}{a^2}$

3. $\dfrac{5 \times 5^4}{5^8}$

4. $\dfrac{t^8}{t^2 \times t^3}$

5. $\dfrac{7}{7^2 \times 7^4}$

6. $\dfrac{m^7 \times m}{m^6}$

7. $\dfrac{4^9 \times 4^{12}}{4^{10} \times 4^3}$

8. $\dfrac{u^8 \times u^{14}}{u^6 \times u^5}$

9. $\dfrac{n^{16} \times n^8}{n \times n^{20}}$

10. $\dfrac{c^6 \times c^2 \times c^3}{c^5 \times c^4}$

11. $\dfrac{9^2 \times 9 \times 9^6}{9^4}$

12. $\dfrac{7^6 \times 7^2 \times 7^4}{7 \times 7^6}$

B Express each of the following as a single power, *then* find its value:

1. $\dfrac{3^5 \times 3^4}{3^7}$

2. $\dfrac{5^6 \times 5^2}{5^6}$

3. $\dfrac{3^7 \times 3^2}{3^6}$

4. $\dfrac{7^5 \times 7^6}{7^9}$

5. $\dfrac{2^8 \times 2^7}{2^{11}}$

6. $\dfrac{5^4 \times 5^6}{5^7}$

7. $\dfrac{8 \times 8^6}{8^5}$

8. $\dfrac{2^4 \times 2^9}{2^7}$

9. $\dfrac{2^2 \times 2^3}{2^8}$

10. $\dfrac{6^3 \times 6^4}{6 \times 6^8}$

11. $\dfrac{5^2 \times 5^4}{5^8}$

12. $\dfrac{10 \times 10^3}{10^3 \times 10^5}$

Exercise 12

A Simplify the following, leaving your answers in index form:

1. $2^3 \div 2^3$

2. $4^2 \div 4^2$

3. $10^4 \div 10^4$

4. $5^3 \div 5^3$

5. $\dfrac{6^2}{6^2}$

6. $\dfrac{3^4}{3^4}$

7. $12^2 \div 12^2$

8. $2^9 \div 2^9$

9. $x^7 \div x^7$

10. h^5/h^5

B Calculate the answers to the following:

1. $2^3 \div 2^3$

2. $4^2 \div 4^2$

3. $10^4 \div 10^4$

4. $5^3 \div 5^3$

82

5. $\dfrac{6^2}{6^2}$

7. $12^2 \div 12^2$

8. $2^9 \div 2^9$

6. $\dfrac{3^4}{3^4}$

C Compare the answers to the first eight questions in part A with the answers to the eight questions in part B. Write what you notice.

D Write the value of:

1. 7^0	**3.** 10^0	**5.** 9^0	**7.** x^0	**9.** p^0
2. 3^0	**4.** 8^0	**6.** 12^0	**8.** y^0	**10.** a^0

Exercise 13

A Answer the following in two different ways as in the example:

e.g. $a^2 \div a^5 = \underline{\underline{a^{-3}}}$

$$a^2 \div a^5 = \frac{a^2}{a^5} = \frac{a \times a}{a \times a \times a \times a \times a} = \frac{1}{\underline{\underline{a^3}}}$$

so $a^{-3} = \dfrac{1}{a^3}$

1. $x^3 \div x^7$ **3.** $p^5 \div p^6$ **5.** $7^6 \div 7^8$

2. $d^4 \div d^6$ **4.** $8^3 \div 8^9$ **6.** $e^2 \div e^4$

B Find the value of:

1. (a) 2^3	(b) 2^{-3}		**4.** (a) 3^4	(b) 3^{-4}
2. (a) 3^2	(b) 3^{-2}		**5.** (a) 10^6	(b) 10^{-6}
3. (a) 5^2	(b) 5^{-2}		**6.** (a) 9^2	(b) 9^{-2}

C Find the value of:

1. 8^{-2}	**4.** 5^{-3}	**7.** 10^{-5}	**10.** 2^{-8}
2. 2^{-4}	**5.** 6^{-2}	**8.** 10^{-8}	**11.** 3^{-6}
3. 3^{-3}	**6.** 10^{-4}	**9.** 2^{-5}	**12.** 20^{-3}

D Copy and complete these, replacing each question mark with the correct symbol ($<$, $>$ or $=$):

1. 10^5 ? 10^7

2. 10^{-5} ? 10^{-7}

3. 1000 ? 10^{-3}

4. 10^{-4} ? $\dfrac{1}{1000}$

5. 10^{-6} ? $0.000\,001$

6. 10^4 ? $100\,000$

7. 0.001 ? $\dfrac{1}{100}$

8. 10^{-7} ? 10^7

9. $\dfrac{1}{10\,000}$? 10^4

10. $\dfrac{1}{10\,000}$? 10^{-4}

11. 0.0001 ? 10^{-3}

12. 0.0001 ? 10^{-4}

13. 10^{-3} ? 0.001

14. $\dfrac{1}{10}$? 10^{-1}

15. 10^{-5} ? $\dfrac{1}{100\,000}$

16. 10^{-3} ? 10^{-4}

Standard Form

Exercise 14

A Write each of the following as numbers without indices:

1. 8.3×10^2
2. 4.6×10^4
3. 3.95×10^3
4. 1.86×10^4
5. 1.04×10^2
6. 9.77×10^5
7. 6×10^4
8. 7.02×10^6
9. 9×10^6
10. 3.8×10^8
11. 2.1×10^7
12. 5.478×10^9

B Write in standard form:

1. 528
2. 7193
3. 8240
4. 7900
5. 61 000
6. 350 000
7. 400 000
8. 6 500 000
9. 21
10. 140 000 000
11. 26 400 000
12. 97 630

C Write the numbers in standard form:

1. The Nile is about 6700 km long.

2. The Moon has a diameter of almost 3500 km.

3. The total area of water in the UK is about 3070 km².

4. Lake Superior (in Canada and the USA) covers an area of about 82 400 km².

5. The Tonga–Kermadec Trench in the South Pacific is about 10 850 m deep at its deepest point.

6. The UK has a land area of about 240 000 km².

7. Canada has a land area of about 9 220 000 km².

8. The distance of the Moon from the Earth is about 380 000 km.

9. The circumference of Jupiter at its equator is almost 447 000 km.

10. Earth is about 150 000 000 km from the Sun.

11. The mean distance of Mars from the Sun is about 228 million kilometres.

12. It is estimated that the population of the UK in the year 2000 AD will be about 62.8 million.

Exercise 15   M

A Copy and complete:

	Standard form	Ordinary number
e.g.	4.2×10^{-4}	0.000 42
1.	6.3×10^{-3}	
2.	3.18×10^{-3}	
3.		0.000 506
4.	9.77×10^{-4}	
5.		0.000 013
6.		0.000 008 451
7.	2.5×10^{-6}	
8.		0.81
9.	5.39×10^{-1}	
10.		0.078

B Write each of the following as a number without an index:

e.g. $3.87 \times 10^{-3} = \underline{0.003\,87}$

1. 2.92×10^{-2}

2. 5.12×10^{-3}

3. 9.64×10^{-4}

4. 1.47×10^{-2}

5. 4.8×10^{-4}

6. 3.51×10^{-1}

7. 2.7×10^{-1}

8. 6.08×10^{-5}

9. 8.132×10^{-2}

10. 7.654×10^{-4}

11. 1.9×10^{-3}

12. 7×10^{-3}

13. 6.99×10^{-6}

14. 4.107×10^{-2}

15. 9.711×10^{-5}

16. 2.04×10^{-4}

17. 3.19×10^{-2}

18. 3.19×10^{-3}

19. 5.016×10^{-1}

20. 5.016×10^{-4}

C Write in standard form:

1. 0.0435

2. 0.000612

3. 0.0054

4. 0.000029

5. 0.08716

6. 0.41

7. 0.0008

8. 0.00654

9. 0.098

10. 0.3429

11. 0.00000023

12. 0.0000719

13. 0.00853

14. 0.000496

15. 0.99

16. 0.603

17. 0.287

18. 0.00287

19. 0.06128

20. 0.00006128

Exercise 16

Write the numbers in standard form:

1. A sheet of silver foil is 0.0002 mm thick.

2. A cube has edges measuring 5 mm. Its volume is 125 mm³ which is 0.125 cm³.

3. The length of a flea is about 0.25 cm which is 0.0025 m.

4. A human blood cell has a diameter of about $0.000\,007$ m.

5. The wavelength of green light is $0.000\,000\,5$ m.

6. The coefficient of linear expansion of ordinary glass is $0.000\,085$ (per degree Celsius).

7. The diameter of a hydrogen atom is $0.000\,000\,000\,106$ m.

8. A robin has a mass of about 0.02 kg.

Exercise 17

Copy these, but replace each question mark with $<$ or $>$ to make each statement correct:

1. $4.6 \quad \times 10^3 \quad \boxed{?} \quad 2.9 \quad \times 10^4$

2. $3.8 \quad \times 10^2 \quad \boxed{?} \quad 6.4 \times 10^{-3}$

3. $8.72 \times 10^{-2} \quad \boxed{?} \quad 7.35 \times 10^{-3}$

4. $5.4 \quad \times 10^{-4} \quad \boxed{?} \quad 4.81 \times 10^{-3}$

5. $9.82 \times 10^{-5} \quad \boxed{?} \quad 9.9 \quad \times 10^{-4}$

6. $1.7 \quad \times 10^4 \quad \boxed{?} \quad 3.21 \times 10^{-5}$

7. $2.91 \times 10^{-4} \quad \boxed{?} \quad 3.21 \times 10^{-5}$

8. $3.06 \times 10^{-1} \quad \boxed{?} \quad 4.12 \times 10^{-1}$

9. $4.64 \times 10^{-3} \quad \boxed{?} \quad 3.989 \times 10^{-3}$

10. $8.251 \times 10^{-5} \quad \boxed{?} \quad 7.6 \quad \times 10^{-5}$

11. $6.49 \times 10^{-1} \quad \boxed{?} \quad 5.8 \quad \times 10^{-1}$

12. $1.32 \times 10^{-3} \quad \boxed{?} \quad 2.5 \quad \times 10^{-3}$

13. $4.97 \times 10^{-6} \quad \boxed{?} \quad 5.01 \times 10^{-6}$

14. $1.45 \times 10^{-3} \quad \boxed{?} \quad 1.83 \times 10^{-4}$

15. $9.1 \quad \times 10^{-5} \quad \boxed{?} \quad 9.08 \times 10^{-6}$

16. $5.37 \times 10^{-4} \quad \boxed{?} \quad 5.41 \times 10^{-3}$

17. $7.6 \quad \times 10^{-2} \quad \boxed{?} \quad 7.4 \quad \times 10^{-2}$

18. $2.07 \times 10^{-5} \quad \boxed{?} \quad 2.13 \times 10^{-5}$

19. $6.2 \quad \times 10^{-7} \quad \boxed{?} \quad 6.187 \times 10^{-7}$

20. $8.3 \quad \times 10^{-3} \quad \boxed{?} \quad 8.29 \times 10^{-3}$

Square Roots

Exercise 18

A Estimate the following square roots by giving two numbers, to one significant figure, between which the square roots lie:

e.g. 1 $\sqrt{46}$ lies between 6 and 7.

e.g. 2 $\sqrt{7000}$ lies between 80 and 90.

1. $\sqrt{20}$	**8.** $\sqrt{820}$	**15.** $\sqrt{6.93}$
2. $\sqrt{90}$	**9.** $\sqrt{1200}$	**16.** $\sqrt{70}$
3. $\sqrt{31}$	**10.** $\sqrt{2800}$	**17.** $\sqrt{7}$
4. $\sqrt{11}$	**11.** $\sqrt{8000}$	**18.** $\sqrt{700}$
5. $\sqrt{75}$	**12.** $\sqrt{68.9}$	**19.** $\sqrt{0.7}$
6. $\sqrt{130}$	**13.** $\sqrt{5.64}$	**20.** $\sqrt{0.07}$
7. $\sqrt{250}$	**14.** $\sqrt{693}$	**21.** $\sqrt{0.007}$

B Answer these using a calculator. Give each answer correct to three significant figures.

1. $\sqrt{1.5}$	**12.** $\sqrt{80}$	**23.** $\sqrt{7.63}$
2. $\sqrt{15}$	**13.** $\sqrt{8}$	**24.** $\sqrt{0.763}$
3. $\sqrt{150}$	**14.** $\sqrt{0.8}$	**25.** $\sqrt{816}$
4. $\sqrt{1500}$	**15.** $\sqrt{0.08}$	**26.** $\sqrt{19.74}$
5. $\sqrt{15\,000}$	**16.** $\sqrt{0.008}$	**27.** $\sqrt{5000}$
6. $\sqrt{0.15}$	**17.** $\sqrt{21}$	**28.** $\sqrt{350}$
7. $\sqrt{0.015}$	**18.** $\sqrt{210}$	**29.** $\sqrt{72}$
8. $\sqrt{0.0015}$	**19.** $\sqrt{2100}$	**30.** $\sqrt{397\,100}$
9. $\sqrt{80\,000}$	**20.** $\sqrt{7630}$	**31.** $\sqrt{48\,300}$
10. $\sqrt{8000}$	**21.** $\sqrt{763}$	**32.** $\sqrt{916.5}$
11. $\sqrt{800}$	**22.** $\sqrt{76.3}$	**33.** $\sqrt{89.07}$

$\sqrt{28}$ lies between 5 and 6.

Now $\sqrt{25} = 5$ and $\sqrt{36} = 6$.

28 is nearer to 25 than to 36 and it so happens that $\sqrt{28}$ is nearer to $\sqrt{25}$ than to $\sqrt{36}$ ($\sqrt{28} = 5.29$ to 3 s.f.).

Is it always true that if a number lies between two consecutive square numbers its square root is closer to the square root of the nearer square number than to the square root of the other square number?

Revision Exercises I to VII

1. $A = \{3, 13, 23, 33, 43\}$ $B = \{$prime numbers less than 50$\}$
 (a) List the numbers in both sets A and B.
 (b) List the numbers in A or B or both.

2. $P = \{$multiples of 3$\}$ $Q = \{$multiples of 5$\}$
 (a) List six numbers that are in both P and Q.
 (b) Describe the set of numbers obtained in part (a).

3. In the Venn diagram if
 $B = \{$blonds$\}$ and
 $L = \{$people with blue eyes$\}$.
 Copy the diagram and shade the
 region that shows the set of blue-
 eyed blonds.

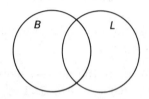

4. $X = \{5, 1, 3, 8, 2\}$ $Y = \{2, 4, 5, 9, 1\}$
 List the members of: (a) both X and Y, (b) X or Y or both.

5. List the elements that are:
 (a) in set M,
 (b) in both M and N,
 (c) not in set N,
 (d) in set M or N or both,
 (e) not in both M and N.

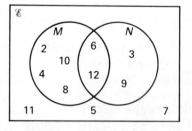

6. The Venn diagram shows the number of elements in a set.
 How many elements are
 there that are:
 (a) in set G?
 (b) not in set F?
 (c) in F or G or both?
 (d) in both F and G?
 (e) not in set G?

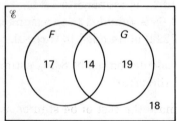

7. If $\mathscr{E}$ = {natural numbers that are less than 40}
P = {prime numbers}
F = {factors of 20}, find:
(a) the number of members of set P,
(b) the number of members that are not in set F,
(c) the members that are in both P and F,
(d) the number of members that are in P or F or both.

8. In a hotel, 76 rooms have a bath, 70 rooms have a TV, 17 rooms have no bath and 9 rooms have a TV but no bath. Draw a Venn diagram and use it to find:
(a) the number of rooms with a bath and TV,
(b) the total number of rooms in the hotel,
(c) the number of rooms that do not have a TV.

Revision Exercise II

1. A box of apples weighs 17.8 kg. Write this to the nearest kilogram.

2. Write to the nearest metre:
(a) 3.4 m   (b) 6.71 m   (c) 82.45 m   (d) 4621 mm

3. Write 8297 g correct to the nearest 100 g.

4. Round 5.476 correct to two decimal places.

5. Round 24.719 correct to three significant figures.

6. Which is the more accurate measurement:
(a) 6 cm or 6.3 cm?   (c) 5.72 kg or 2.6 kg?
(b) 2.9 ℓ or 4.67 ℓ?   (d) 3.01 m or 3.1 m?

7. Calculate the error if:
(a) the actual length = 7.36 m and the recorded length = 7.4 m,
(b) 12.7 kg is rounded to 13 kg,
(c) 1.84 ℓ is rounded to 1.8 ℓ,
(d) 4.68 is rounded to two significant figures,
(e) 72.89 is rounded to three significant figures.

8. The calculation 3.6 × 8.3 is estimated from 4 × 8.
Find the error.

9. Estimate the cost of 54 stamps at 17 p each.

Revision Exercise III

1. Work out:
   (a) 297 + 75
   (b) 7032 − 4367
   (c) 6 × 14
   (d) 34 × 18
   (e) 1800 ÷ 6
   (f) 3822 ÷ 7
   (g) 4160 ÷ 20
   (h) 1035 ÷ 23

2. Find the difference between 16.2 and 2.74.

3. The value of 41.007 × 100 is:
   A. 41.007 00   B. 417   C. 4100.7   D. 0.410 07
   (Choose the correct answer from above.)

4. Work out:
   (a) 3.7 + 0.69
   (b) 4.87 + 29.6 + 0.2
   (c) 8.7 − 29.5
   (d) 81.4 − 47.86
   (e) 6.24 × 3
   (f) 42.8 × 2.1
   (g) 14.5 ÷ 5
   (h) 4 ÷ 0.02

5. A number, when divided by 9, gives an answer of 7 and a remainder of 4. Find the number.

6. Find the value of:
   (a) 36 − (19 + 8)
   (b) 7 + 2 × 9

7. Which is bigger, 0.064 or 0.46?

8. If 38 × 26 = 988, find the value of:
   (a) 3.8 × 26
   (b) 3.8 × 2.6
   (c) 380 × 0.026
   (d) 98.8 ÷ 2.6
   (e) 9880 ÷ 38
   (f) 98.8 ÷ 0.38

9. Find 54.73 ÷ 5 correct to two decimal places.

10. Find 24.8 ÷ 1.7 correct to three significant figures.

11. A car travels 179 miles on 8 gal of petrol.
    Find the number of miles per gallon correct to one decimal place.

12. After entering a number on a calculator, I divided by 3.1, added 7.4, multiplied the answer by 2.3 then subtracted 4.68. The answer obtained was 25.22. Find the number I started with.

13. Answer these. Where possible, simplify your answers.

(a) $2\dfrac{3}{4} + 3\dfrac{2}{3}$ (c) $4\dfrac{1}{4} - 3\dfrac{5}{6}$ (e) $\dfrac{5}{8} \div \dfrac{3}{4}$

(b) $7\dfrac{3}{5} - 4\dfrac{1}{2}$ (d) $\dfrac{3}{8} \times \dfrac{6}{7}$ (f) $5\dfrac{1}{4} \div \dfrac{7}{8}$

14. A motorist had $24\,\ell$ of oil in a drum. How many litres were left after $\frac{3}{4}$ of the oil was used?

15. Change to decimals:

(a) $\dfrac{2}{5}$ (b) $\dfrac{5}{8}$ (c) $\dfrac{21}{25}$ (d) $\dfrac{3}{16}$

16. Change to decimals giving each answer correct to three decimal places:

(a) $\dfrac{6}{7}$ (b) $\dfrac{5}{6}$ (c) $\dfrac{7}{12}$

17. Mrs Magill withdrew £47 from her bank account each month for 5 months. Work out her total withdrawal.

18. Find the two missing numbers in the following sequence:
    $15,\ 8,\ \boxed{?},\ ^{-}6,\ ^{-}13,\ \boxed{?},\ ^{-}27,\ ^{-}34.$

19. Give the first six terms of each sequence using:
    (a) the rule 'double then add 7' on the numbers $\{3, 4, 5, 6, 7, 8\}$,
    (b) $T_n = 5n + 2$

20. Find a rule or a formula for T_n for the sequence:
    $4, 7, 10, 13, 16, 19, 22, \ldots$

Revision Exercise IV ▬▬▬▬▬▬▬▬▬▬▬▬▬▬▬▬

1. Convert to metres:
   (a) $482\,\text{cm}$ (b) $9130\,\text{mm}$ (c) $8.06\,\text{km}$

2. Convert to centimetres:
   (a) $68\,\text{mm}$ (b) $2.7\,\text{m}$ (c) $1.6\,\text{km}$

3. Convert $3980\,\text{m}$ to kilometres.

4. Convert to grams:
   (*a*) 8.65 kg   (*b*) 0.46 kg   (*c*) 64 500 mg

5. Convert to kilograms:
   (*a*) 5800 g   (*b*) 900 g   (*c*) 2.14 t

6. Convert to litres:
   (*a*) 4550 mℓ   (*b*) 13 720 mℓ   (*c*) 650 mℓ

7. Convert to millilitres:
   (*a*) 9.33 ℓ   (*b*) 12.6 ℓ   (*c*) 0.45 ℓ

8. Convert to tonnes:
   (*a*) 3700 kg   (*b*) 59 400 kg   (*c*) 320 kg

9. A piece of curtain material is 6.2 m long. What length remains if 2.7 m is cut off?

10. A picture frame needs a length of 0.92 m of wood. What total length is needed to make seven of these frames?

Revision Exercise V

1. The table shows the monthly repayments on a mortgage:

Monthly Repayments at a Mortgage Rate of 8.52%

Mortgage	Length of mortgage (years)			
	10	15	20	25
£10 000	£127.10	£100.50	£88.20	£81.60
£15 000	£190.65	£150.75	£132.30	£122.40
£20 000	£254.20	£201	£176.40	£163.20
£25 000	£317.75	£251.25	£220.50	£204

   (*a*) What is the monthly repayment on £20 000 over 25 years?
   (*b*) What is the monthly repayment on £15 000 over 10 years?
   (*c*) What is the monthly repayment on £10 000 over 20 years?

(d) What would be the monthly repayment on £5000 over 20 years?

(e) What would be the monthly repayment on £10 000 over 15 years?

(f) What would be the monthly repayment on £1000 over 15 years?

(g) What would be the monthly repayment on £12 000 over 15 years?

(h) What would be the monthly repayment on £14 000 over 25 years?

(i) On a 15-year mortgage, what is the total annual repayment on a £20 000 loan?

(j) On a 15-year mortgage, what is the total amount repaid over 15 years on a £20 000 loan?

(k) By how many times is the repayment on a 15-year mortgage of £20 000 bigger than the loan?

2. If mortgage rates are increased from 8.52% net to 9.94% net, then the monthly repayments are as follows:

Monthly Repayments at a Mortgage Rate of 9.94%

Mortgage	Length of mortgage (years)			
	10	15	20	25
£10 000	£135.30	£109.20	£97.50	£91.40
£15 000	£202.95	£163.80	£146.25	£137.10
£20 000	£270.60	£218.40	£195	£182.80
£25 000	£338.25	£273	£243.75	£228.50

(a) What is the extra monthly payment on a £10 000 mortgage over 10 years?

(b) On a 15-year mortgage, what is the total annual repayment on a £20 000 loan?

(c) On a 15-year mortgage, what is the total amount repaid over 15 years on a £20 000 loan?

(d) How much extra is paid on a £20 000 loan over 15 years when compared with the previous rate?

3. Some cars have rev counters (tachometers).

On the rev counter illustrated, the number 4 stands for 4000 rev/min.

What is the reading shown?

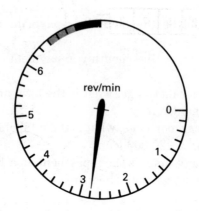

4. Give the readings on the following electricity meters in kWh. Ignore the $\frac{1}{10}$ kWh digit.

(a)

| 0 | 8 | 1 | 5 | 7 | $\frac{1}{10}$ 5 | kWh |

(b)

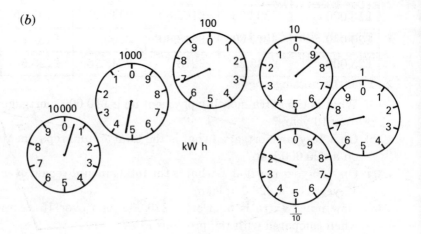

5. Give the readings on the following meters in hundreds of cubic feet of gas (ignore the tens and units digits):

(a)

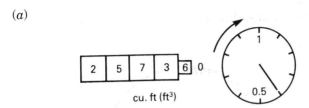

cu. ft (ft³)

(b)

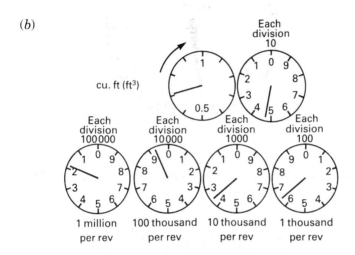

cu. ft (ft³)

Revision Exercise VI

1. Find the sum of the interior angles of:
   (a) a quadrilateral,   (b) a decagon (a 10-sided polygon).

2. Find the missing angles:

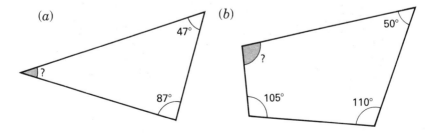

(a)

(b)

97

3. Two angles of a triangle measure 74° and 29°. Find the third angle.

4. Calculate each interior angle of a regular octagon.

5. Each interior angle of a regular polygon is 144°. How many sides has it got?

6. A regular octagon is shown.
O is the centre.
Calculate angle FOE.

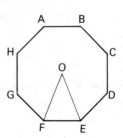

7. Squares tessellate on their own.
Equilateral triangles tessellate on their own.
Name another regular polygon that tessellates on its own.

8. Name two polygons that tessellate together.

Revision Exercise VII

1. Here is a set of numbers:
$\{4, \ ^-2, 3, 0, \sqrt{5}, 7.3, \ ^-1.6, 1\frac{2}{3}, \sqrt{9}, \ ^-3.7, \ ^-\sqrt{11}, \frac{3}{5}, \ ^-7\}$
From the set above, list the set of:
(*a*) natural numbers,   (*e*) whole numbers,
(*b*) integers,   (*f*) negative integers,
(*c*) rational numbers,   (*g*) non-positive integers,
(*d*) positive integers,   (*h*) irrational numbers.

2. What is the value of each of the following?
(*a*) 4^2 (*b*) 13^2 (*c*) 2^6 (*d*) 5^5

3. Use a calculator to find 4.87^2 correct to three significant figures.

4. If $y = 3x^2$, find y when x equals:
(*a*) 4   (*b*) $^-2$

5. If $a = 3$, find the value of:
(*a*) $4a + 2a$ (*b*) $4a \times 2a$ (*c*) $6a$ (*d*) $8a^2$

6. Simplify, leaving answers in index form:

(a) $3d \times 5d$ (d) $h^7 \times h^5$ (f) $\dfrac{t^8}{t^3}$

(b) $2n^3 \times 3n^2$ (e) $\dfrac{x^6}{x^2}$ (g) $5m^2v^3 \times 3mv^4$

(c) $6k \times 2k^4$

7. Simplify by expressing as a single power:

(a) $\dfrac{x^{10}}{x^2}$ (b) $\dfrac{8^{12}}{8^3}$ (c) $\dfrac{m^7 \times m^{12}}{m^5}$ (d) $\dfrac{9^6 \times 9^7}{9^3}$

8. Express as a single power then find the value of:

(a) $\dfrac{7^7}{7^4}$ (b) $\dfrac{8^6}{8^8}$ (c) $\dfrac{2^9 \times 2^4}{2^8}$ (d) $\dfrac{10^8 \times 10}{10^{12}}$

9. Find the value of:

(a) 7^{-2} (b) 4^{-3}

10. Express in standard form ($a \times 10^n$ where $1 \leqslant a < 10$ and n is an integer):

(a) 4290 (b) 306 000 (c) 0.04 (d) 0.000 056

11. Between which two consecutive whole numbers does $\sqrt{135}$ lie?

12. Give two numbers, each correct to one significant figure, between which these square roots lie:

(a) $\sqrt{59}$ (b) $\sqrt{600}$ (c) $\sqrt{8.64}$ (d) $\sqrt{0.005}$

13. Use a calculator to find, correct to three significant figures:

(a) $\sqrt{42}$ (b) $\sqrt{420}$ (c) $\sqrt{4.2}$ (d) $\sqrt{42\,000}$

14. Use a calculator to find, correct to three significant figures:

(a) $\sqrt{8.9}$ (b) $\sqrt{407}$ (c) $\sqrt{0.255}$ (d) $\sqrt{0.0675}$

8 Brackets and Factorising

Exercise 1

Multiply out:

1. $2(x + 5)$
2. $5(t - 2)$
3. $4(m + 3)$
4. $3(q + 5)$
5. $6(u - 4)$
6. $9(n - 3)$
7. $8(a + 6)$
8. $2(l - 9)$
9. $3(3 + f)$
10. $5(4 - c)$

11. $7(1 + g)$
12. $6(5 - k)$
13. $5(v + 1)$
14. $4(z - 2)$
15. $8(3 - y)$
16. $2(3x + 1)$
17. $3(3b + 4)$
18. $5(4s + 3)$
19. $2(7w - 5)$
20. $4(2c - 3)$

21. $6(3d - 2)$
22. $3(9b + 7)$
23. $2(7a + 6)$
24. $9(3e - 2)$
25. $5(5z + 5)$
26. $4(5e - 4)$
27. $8(2n - 4)$
28. $7(1 + 2h)$
29. $6(3 - 2p)$
30. $7(6 + 3v)$

Exercise 2

Multiply out:

1. $^-2(x - 3)$
2. $^-3(a + 5)$
3. $^-6(f - 1)$
4. $^-4(z - 6)$
5. $^-7(t + 2)$
6. $^-5(c - 5)$
7. $^-2(k + 8)$

8. $^-9(2 + e)$
9. $^-3(2 + q)$
10. $^-4(7 - m)$
11. $^-8(2b - 3)$
12. $^-5(3g + 4)$
13. $^-2(7n + 6)$
14. $^-3(3r - 3)$

15. $^-6(3l + 5)$
16. $^-5(7s + 6)$
17. $^-4(4d + 2)$
18. $^-3(1 - 6p)$
19. $^-7(3 + 2p)$
20. $^-8(4 - 7u)$

Exercise 3

If $A = (x - 4)$, $B = (x - 6)$, $C = (2x + 5)$, $D = (3x - 2)$, $E = (5 - x)$, $F = (3 - 4x)$, $G = (2x - 7)$ and $H = (4x - 1)$, find and simplify by multiplying out:

1. $3B$
2. $2A$
3. $5A$
4. $7B$
5. $3C$
6. $5D$
7. $4G$
8. $6H$

9. $3E$	12. $3H$	15. ^-E	18. ^-4H
10. $2F$	13. ^-A	16. ^-F	19. ^-2E
11. $7G$	14. ^-D	17. ^-3G	20. ^-3F

Exercise 4

Factorise:

1. $2u + 6$	9. $18w + 27$	17. $14 + 2m$
2. $3k + 12$	10. $21b + 35$	18. $12 - 4t$
3. $5h - 5$	11. $24x - 20$	19. $35 - 5a$
4. $4n - 16$	12. $15g - 27$	20. $15 + 6e$
5. $6p + 36$	13. $10s + 25$	21. $6 + 30k$
6. $6d + 4$	14. $16l + 24$	22. $2 - 18r$
7. $12f + 9$	15. $12v - 10$	23. $28 - 21y$
8. $12q - 6$	16. $24c + 18$	24. $9 + 15z$

Exercise 5

Find the value of the following by factorising first:

1. $7 \times 6 + 7 \times 4$	9. $8 \times 6.3 + 8 \times 3.7$
2. $9 \times 34 - 9 \times 24$	10. $5 \times 31.9 - 5 \times 21.9$
3. $3 \times 29 + 3 \times 31$	11. $1.6 \times 19 + 19 \times 8.4$
4. $6 \times 47 + 6 \times 23$	12. $6.8 \times 4.5 + 6.8 \times 5.5$
5. $17 \times 8 + 8 \times 33$	13. $4.7 \times 51.8 + 48.2 \times 4.7$
6. $43 \times 65 - 55 \times 43$	14. $66.3 \times 7.79 + 66.3 \times 2.21$
7. $57 \times 38 + 57 \times 62$	15. $51.6 \times 18.3 + 51.6 \times 81.7$
8. $81 \times 72 - 61 \times 72$	16. $3.142 \times 9 - 3.142 \times 4$

Exercise 6

Simplify the following. If an expression cannot be simplified, write 'NO SIMPLER FORM'.

1. $n + n + n + n$	5. $6 + 3 + 5 + 7$
2. $2t + 6t$	6. $x + x + y + x + y$
3. $4c + 3c + 5c$	7. $4g + 3h + 2g + 6h$
4. $9z + 4z + z + 5z$	8. $7e + 2e + 6f + 3e + f$

9. $6u - 2u$
10. $15q - 9q$
11. $13k - 2k - k$
12. $16m + 8m - 7m$
13. $9v - 6v + 3v$
14. $18r - 12r + 14r$
15. $9d - 8d + 3d - d - 2d$
16. $19a + 3a - 12a - 4a + a$
17. $19 + 3 - 12 - 4 + 1$
18. $3n - 2n + 5n - n + 6n$
19. $4p + 3s - 2p + 2s$

20. $9w - 7w + 3b - b$
21. $19d + 7e$
22. $4z + 9 - 3z - 7 + 8z$
23. $11v + 5c - 4c - 3v$
24. $4b + 3q + 3b - q - 7b$
25. $8f + 2g - 3f + 6h$
26. $14k + 3l + 5m$
27. $9s - 3s + t + 6u + 4t$
28. $4x + 8y + 3 - 2x + 6 - y$
29. $7w + 3a - a - 3w - 2a$
30. $23p + r - 2p + 4l - r$

Exercise 7

Simplify:

1. (a) $3 + 9 + 4$
(b) $3a + 9a + 4a$

2. (a) $10 - 4 + 3$
(b) $10b - 4b + 3b$

3. (a) $4 - 7$
(b) $4c - 7c$

4. (a) $6 - 11$
(b) $6d - 11d$

5. (a) $7 - 9 + 6$
(b) $7e - 9e + 6e$

6. (a) $1 - 8 + 12$
(b) $f - 8f + 12f$

7. (a) $3 - 12 + 2$
(b) $3g - 12g + 2g$

8. (a) $9 - 6 - 7$
(b) $9h - 6h - 7h$

9. (a) $4 - 11 - 7$
(b) $4k - 11k - 7k$

10. (a) $6 - 8 + 1 - 13$
(b) $6l - 8l + l - 13l$

Exercise 8

A Find the value of:

1 (a) $9 + 2 - 3 + 4$
(b) $9 + 2 + 4 - 3$

2. (a) $4 - 2 + 6 - 5$
(b) $4 + 6 - 2 - 5$

3. (a) $3 - 7 + 10 + 2$
(b) $3 + 10 + 2 - 7$

4. (a) $2 - 13 - 11 + 15$
(b) $2 + 15 - 13 - 11$

5. (a) $10 - 4 - 19 + 3$
(b) $10 + 3 - 4 - 19$

6. (a) $15 - 19 + 7 - 3 + 8$
(b) $15 + 7 + 8 - 19 - 3$

7. (a) $4 - 21 - 5 + 18 - 3 + 9$
     (b) $4 + 18 + 9 - 21 - 5 - 3$

8. (a) $^-3 + 7 - 4 + 9 - 2$
     (b) $7 + 9 - 3 - 4 - 2$

9. (a) $^-8 - 6 + 5 - 13 + 4$
     (b) $5 + 4 - 8 - 6 - 13$

10. (a) $^-5 - 11 + 3 - 2 + 9 + 14$
      (b) $3 + 9 + 14 - 5 - 11 - 2$

B Simplify:

1. (a) $4m - 2m + 6m$
     (b) $4m + 6m - 2m$

2. (a) $9n - 4n + 2n - 3n$
     (b) $9n + 2n - 4n - 3n$

3. (a) $12p + 3p - 9p + 8p$
     (b) $12p + 3p + 8p - 9p$

4. (a) $6q - 8q - 3q + 9q$
     (b) $6q + 9q - 8q - 3q$

5. (a) $7r - 2r + 5r - 8r + r$
     (b) $7r + 5r + r - 2r - 8r$

6. (a) $16s - s - 9s + 4s - 3s$
     (b) $16s + 4s - s - 9s - 3s$

7. (a) $2t - 7t + t + 8t - 11t$
     (b) $2t + t + 8t - 7t - 11t$

8. (a) $5u - 9u - u - 3u + 8u$
     (b) $5u + 8u - 9u - u - 3u$

Exercise 9

Simplify:

1. $4v - 3w + 6v + 5w$
2. $9x + 4y - 7y + 2x$
3. $12z + 3a - 6z - 7a$
4. $b - 6c - 2c + 4b$
5. $3d - 4e + 7e - 5d$
6. $7f + g - 5g - 2f$
7. $h + 3k - 4h - 2k$
8. $4l - 2l - 9m - 6m$
9. $n - 5p + 6p - 8n$
10. $2q - 2r + 9r + 12q$
11. $9s - 7s - t - 4t + 2t$
12. $2w - 8w + v - 3v + 5v$
13. $4u - 3x - 2x + u - 5u$
14. $y - 7z + 3y - 5y + 6z$
15. $11a + 5b - 6a - 5b - 5a$
16. $9c + d - 3c + 3d - 4c$
17. $12e - f - 9e - f - 4e - f$
18. $g + h - 3g + 2h + 6g$
19. $2k - l - 4l + 5k + 5l$
20. $15m - 3n + 11n - 13m - 4m$

Exercise 10

A **1.** If $c = 3$ and $d = 6$, find the value of:
    (*a*) *cd*   (*b*) *dc*

2. If $m = 4$ and $n = 7$, find the value of:
    (*a*) *mn*   (*b*) *nm*

3. If $p = 5$ and $q = {}^{-}3$, find the value of:
    (*a*) *pq*   (*b*) *qp*

4. If $a = {}^{-}2$ and $b = {}^{-}8$, find the value of:
    (*a*) *ab*   (*b*) *ba*

5. If $x = 9$ and $y = 7$, find the value of:
    (*a*) *xy*   (*b*) *yx*

B **1.** If $e = 4$ and $f = 5$, find the value of:
    (*a*) $2ef + 4ef$ (*b*) $6ef$ (*c*) $5fe + ef$

2. If $s = 2$ and $t = 7$, find the value of:
    (*a*) $3st + 5st$ (*b*) $8st$ (*c*) $6ts + 2ts$

3. If $g = {}^{-}2$ and $h = {}^{-}3$, find the value of:
    (*a*) $4gh + 2hg$ (*b*) $6gh$ (*c*) $8gh - 2hg$

4. If $k = 4$ and $l = {}^{-}2$, find the value of:
    (*a*) $6lk + kl - 4kl$ (*b*) $5lk - 2kl$ (*c*) $3kl$

Exercise 11

Where possible, simplify the following. If an expression cannot be simplified, write 'NO SIMPLER FORM'.

1. $5pq + 3pq - 4pq$ **7.** $5xy + 3x - 2xy + 4x$

2. $7mj - 2jm + mj + 5jm$ **8.** $5mn + 8mp - nm - 5pm$

3. $4e + 3fe - 2e + fe - 2fe$ **9.** $3tu + 5u + 7ut - u - 2tu$

4. $5gh + 2g - 2gh - 3hg - g$ **10.** $2vw - vw + 3v + 5wv$

5. $9de + 3fg - 2de + 3gf$ **11.** $5gp + 2ph - 3pg$

6. $4mv + 3wx - vm - 2xw$ **12.** $4ad + 3ac + 5cd$

13. $5un + 3np - 2pn$ **17.** $4klm + 3klm$

14. $ac + 3a + 4ac - a - 2ca$ **18.** $4klm + 2kml$

15. $3n + 2p + 3pn - 2p - np$ **19.** $7pqr - 7qrp$

16. $4d - de + 3ed + 4e + d + 5de$ **20.** $3fed - edf + 4fad$

Exercise 12

A Work out:

1. (a) $5 + (4 + 2)$ (b) $5 + 4 + 2$ (c) $19 - 8 - 3$

2. (a) $13 + (3 + 6)$ (b) $13 + 3 + 6$ (c) $34 - 20 - 5$

3. (a) $12 + (7 - 2)$ (b) $12 + 7 - 2$ (c) $14 - 9 - 5$

4. (a) $21 + (15 - 6)$ (b) $21 + 15 - 6$ (c) $43 - 24 - 9$

5. (a) $19 - (8 + 3)$ (b) $19 - 8 + 3$

6. (a) $34 - (20 + 5)$ (b) $34 - 20 + 5$

7. (a) $14 - (9 - 5)$ (b) $14 - 9 + 5$

8. (a) $43 - (24 - 9)$ (b) $43 - 24 + 9$

B Simplify:

1. $5 + (x + 2)$ **10.** $15 + (17 - 4u)$

2. $9 + (a - 4)$ **11.** $n + (n + 7)$

3. $7 + (5 + d)$ **12.** $4b + (b + 8)$

4. $8 + (6 - y)$ **13.** $w + (3w - 6)$

5. $9 + (3t + 4)$ **14.** $3f + (5f + 10)$

6. $12 + (5h - 7)$ **15.** $5k + (2k - 6)$

7. $21 + (2v - 13)$ **16.** $8z + (9 - 2z)$

8. $7 + (6 + 3p)$ **17.** $12l + (13 - 4l)$

9. $9 + (8 - 2e)$ **18.** $7g + (8g - 19)$

C Simplify:

1. $9 - (i + 3)$

2. $16 - (s + 8)$

3. $14 - (q - 7)$

4. $11 - (r - 11)$

5. $19 - (12 + c)$

6. $13 - (11 - x)$

7. $8 - (8 - t)$

8. $12 - (2n - 5)$

9. $12 - (2n + 5)$

10. $2 - (4r + 2)$

11. $5h - (3h - 7)$

12. $7u - (u + 9)$

13. $10b - (8b + 1)$

14. $4f - (4f - 12)$

15. $16v - (6 - v)$

16. $16 - (6 - v)$

17. $14a - (3 + 9a)$

18. $j - (1 - j)$

19. $18e - (8 - 2e)$

20. $2p - (p - 15)$

Exercise 13

A Simplify:

1. $4 + 3(x + 2)$

2. $8 + 5(x + 1)$

3. $9 + 4(x - 2)$

4. $15 + 2(y - 3)$

5. $19 + 7(y - 2)$

6. $14 + 3(4 + z)$

7. $6 + 2(9 + t)$

8. $1 + 7(2 - m)$

9. $9 + 4(5 - c)$

10. $3b + 2(b + 4)$

11. $6q + 3(q - 1)$

12. $11n + 4(n - 7)$

13. $19 + 2(3g + 5)$

14. $13 + 5(2a - 3)$

15. $15 + 3(4 - w)$

16. $12 + 2(6 - 3u)$

17. $4d + 3(7 + 2d)$

18. $9n + 2(5n - 1)$

19. $7e + 3(4e - 6)$

20. $9p + 5(2p - 1)$

B Simplify:

1. $9 - 2(w + 3)$

2. $8 - 2(e - 4)$

3. $15 - 3(p + 3)$

4. $20 - 5(a + 4)$

5. $25 - 4(x - 1)$

6. $18 - 3(g - 2)$

7. $41 - 2(7 + u)$

8. $23 - 4(5 - c)$

9. $6 - 3(2 - n)$

10. $10 - 6(1 - v)$

11. $4k - 2(k - 9)$

12. $2y - 2(y + 1)$

13. $11 - 5(3d + 2)$

14. $8 - 3(4s - 5)$

15. $16 - 4(6b - 4)$

16. $12 - 3(3 - 3t)$

17. $6f - 2(2f + 3)$

18. $10m - 3(3m - 6)$

19. $16l - 7(2l - 5)$

20. $24 - 4(5l + 5)$

Exercise 14

Simplify:

1. $x^2 + 4x - 2x + 5$

2. $x^2 - 3x + 2x - 6$

3. $x^2 - 5x + 7x - 2$

4. $y^2 + y - 4y - 3$

5. $y^2 - 5y + 5y - 1$

6. $t^2 + 6t - 6t + 4$

7. $t^2 - 9t + 9t - 3$

8. $p^2 + 4p + 4p + 2$

9. $d^2 - 3d - 3d + 9$

10. $a^2 - 7a + 7a - 7$

11. $f^2 + 8f - 8f + 2$

12. $z^2 - z + z - 4$

Exercise 15

Multiply out:

e.g. 1 $x(x + 3)$
$$= \underline{\underline{x^2 + 3x}}$$

e.g. 2 $x(x - 3)$
$$= \underline{\underline{x^2 - 3x}}$$

1. $x(x + 5)$

2. $a(a + 2)$

3. $4(y - 4)$

107

4. $2(z - 2)$ **7.** $n(n + 8)$ **10.** $7(k - 7)$ **13.** $g(g + 9)$

5. $p(p - 6)$ **8.** $u(u + 1)$ **11.** $t(t - 5)$ **14.** $q(q - 10)$

6. $6(p - 6)$ **9.** $k(k - 7)$ **12.** $5(t - 5)$ **15.** $m(m - 8)$

Exercise 16

Multiply out and simplify your answers:

e.g. $x(x - 3) + 5(x - 3)$
$$= x^2 - 3x + 5x - 15$$
$$= \underline{x^2 + 2x - 15}$$

1. $x(x - 2) + 4(x - 2)$ **7.** $h(h - 2) + 2(h + 2)$

2. $c(c + 4) + 2(c - 4)$ **8.** $k(k + 7) + 7(k + 3)$

3. $d(d - 3) + 2(d + 3)$ **9.** $l(l - 9) + 9(l - 4)$

4. $e(e - 3) + 3(e + 3)$ **10.** $m(m - 4) + 4(m - 8)$

5. $f(f + 3) + 3(f - 3)$ **11.** $n(n - 12) + 12(n + 12)$

6. $g(g - 7) + 7(g + 7)$ **12.** $p(p - 12) + 12(p - 12)$

Exercise 17

Multiply out and simplify your answers:

e.g. $x(x + 5) - 3(x - 6)$
$$= x^2 + 5x - 3x + 18$$
$$= \underline{x^2 + 2x + 18}$$

1. $x(x + 4) - 2(x - 5)$ **7.** $v(v + 3) - 5(v + 1)$

2. $q(q + 7) - 5(q + 2)$ **8.** $w(w - 8) - 7(w + 2)$

3. $r(r - 2) - 3(r + 1)$ **9.** $y(y - 4) - 5(y + 6)$

4. $s(s + 6) - 4(s - 5)$ **10.** $z(z + 9) - 6(z - 4)$

5. $t(t - 1) - 2(t - 6)$ **11.** $a(a + 2) - 8(a + 2)$

6. $u(u - 6) - 9(u - 2)$ **12.** $b(b - 5) - 3(b - 8)$

Exercise 18

e.g. $(x + 2)(x + 3)$
$$= x(x + 3) + 2(x + 3)$$
$$= x^2 + 3x + 2x + 6$$
$$= \underline{x^2 + 5x + 6}$$

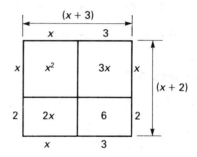

Multiply out and simplify your answers:

1. $(x + 2)(x + 4)$ **7.** $(u + 1)(u + 10)$

2. $(x + 3)(x + 5)$ **8.** $(a + 4)(a + 6)$

3. $(x + 6)(x + 1)$ **9.** $(y + 7)(y + 2)$

4. $(x + 3)(x + 4)$ **10.** $(t + 2)(t + 9)$

5. $(x + 6)(x + 2)$ **11.** $(p + 5)(p + 4)$

6. $(x + 12)(x + 1)$ **12.** $(n + 8)(n + 3)$

Exercise 19

Multiply out and simplify your answers:

1. $(x + 4)(x - 2)$ **7.** $(c + 3)(c - 3)$

2. $(d + 3)(d - 6)$ **8.** $(r + 8)(r - 1)$

3. $(g + 1)(g - 8)$ **9.** $(w + 5)(w - 5)$

4. $(s + 6)(s - 3)$ **10.** $(f + 7)(f - 6)$

5. $(v + 5)(v - 4)$ **11.** $(m + 9)(m - 4)$

6. $(q + 2)(q - 7)$ **12.** $(h + 4)(h - 7)$

Multiply out and simplify your answers:

e.g. $(x - 2)(x + 3)$
$$= x(x + 3) - 2(x + 3)$$
$$= x^2 + 3x - 2x - 6$$
$$= \underline{x^2 + x - 6}$$

1. $(x - 2)(x + 6)$

2. $(x - 3)(x - 2)$

3. $(x - 4)(x - 1)$

4. $(x - 1)(x + 5)$

5. $(x - 6)(x - 2)$

6. $(x - 8)(x + 3)$

7. $(b - 3)(b + 1)$

8. $(z - 7)(z - 2)$

9. $(u - 10)(u - 4)$

10. $(e - 4)(e + 4)$

11. $(m - 6)(m + 3)$

12. $(k - 5)(k - 3)$

9 Solids and Constructions

Solids and their Nets

Here is a drawing of a cube:

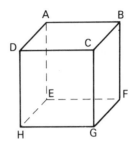

The base EFGH is a square but it has been drawn as a parallelogram:

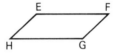

Here is a drawing of a cylinder:

The base of a cylinder is a circle but it is necessary to draw it as an ellipse:

Exercise 1

1. Draw a sketch of a cylinder.

2. Draw a sketch of a cone.

111

Exercise 2

1. Here is a sketch of a triangular-based prism:
   (a) Copy it.
   (b) How many faces has it got?
   (c) How many edges has it got?
   (d) How many vertices has it got?

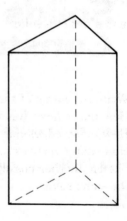

2. Here is a sketch of a tetrahedron (a triangular-based pyramid):

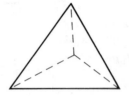

   (a) Copy it.
   (b) How many faces has it got?
   (c) How many edges has it got?
   (d) How many vertices has it got?

Exercise 3

Where possible, write the name of a solid that has exactly:

1. (a) one face
   (b) two faces
   (c) three faces
   (d) four faces
   (e) five faces
   (f) six faces

2. (a) one edge
   (b) two edges
   (c) three edges
   (d) four edges
   (e) five edges
   (f) six edges

3. (a) one vertex
   (b) two vertices
   (c) three vertices
   (d) four vertices
   (e) five vertices
   (f) six vertices

Exercise 4

A Here is a net of a solid:

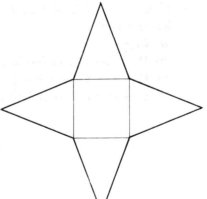

1. Write the name of the solid.
2. How many faces has it got?
3. How many edges has it got?
4. How many vertices has it got?
5. Sketch another possible net of the same solid.

B Draw two different possible nets of a regular tetrahedron.

Exercise 5

1. Here is a net of a cube:

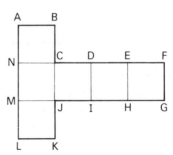

When the net is folded into a cube, to which letters will each of the given points join?

(a) B   (b) I   (c) A   (d) H   (e) F   (f) M

2. Here is another net of a cube:

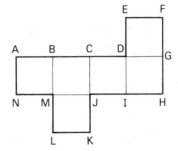

When the net is folded into a cube, to which letters will each of the given points join?

(a) C   (b) I   (c) A   (d) B

3. When the net in question 2 is folded into a cube, three letters join together to meet at one vertex of the cube. Which three letters are they?

Exercise 6

Make two copies of the given net on a piece of card then cut them out:

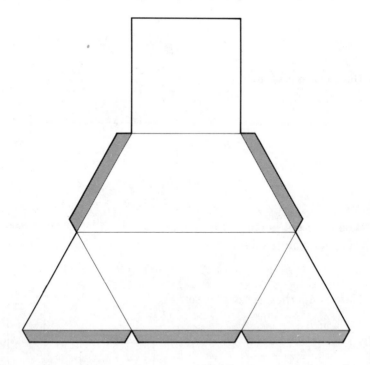

Fold and stick each one to make two identical solids.

Puzzle: Put the two solids together to form a tetrahedron.

1. Draw, then cut out two copies of the given pattern on a piece of card:

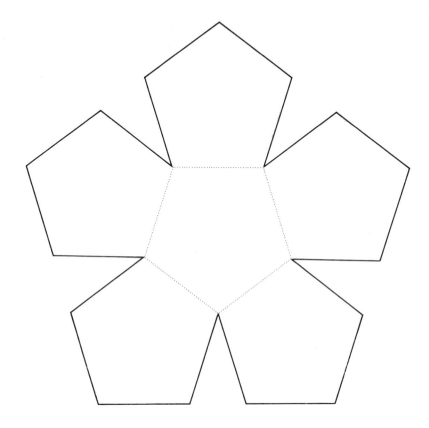

2. Score the edges that have dotted lines.

3. Fold the pentagons downwards along the dotted lines.

4. Flatten out the two pieces again then place them together positioning them as in the diagram on the next page. Ensure that the two parts bend slightly towards each other.

5. Thread an elastic band alternately above and below the vertices of each piece so that the band tends to hold the two pieces together. (The position of the elastic band is shown by the thick lines in the diagram on the next page.)

If you now let go, the shape should pop up into a dodecahedron.

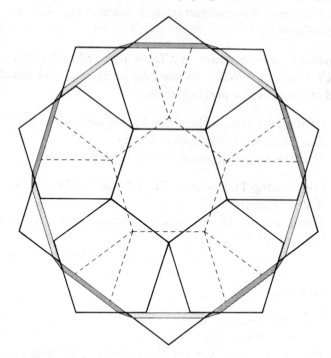

Constructions

Exercise 8

1. Construct triangle ABC such that AB = 55 mm, BC = 45 mm and AC = 50 mm. Measure angle ABC.

2. Construct triangle DEF such that DE = 5 cm, EF = 3 cm and angle E = 56°. How long is DF?

3. (a) Make a full-size drawing of △PQR.
   (b) Measure angle PQR.
   (c) Using a pair of compasses bisect angle PRQ and let the bisector meet PQ at X.
   (d) How long is QX?

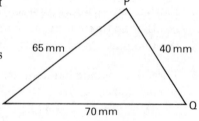

4. Construct △JKL where JK = 50 mm, KL = 60 mm and LJ = 36 mm. By measuring and calculating, find the area of △JKL.

5. Construct parallelogram WXYZ where XY = 55 mm, angle WXY = 44° and WX = 36 mm. By measuring and calculating find the area of the parallelogram.

6. (a) Construct a regular hexagon having sides of length 35 mm. Start with a circle of radius 35 mm.
   (b) Find the length of one of the shortest diagonals.

7. (a) Construct △TUV where TU = 50 mm, UV = 80 mm and TV = 70 mm.
   (b) Using a pair of compasses, construct the perpendicular bisectors of TV and UV.
   (c) Label the point where the perpendicular bisectors meet as O.
   (d) Draw a circle, centre O, with radius OT.
   (e) How long is OT?

8. (a) Draw any triangle.
   (b) Bisect all three sides.
   (c) What do you notice about the perpendicular bisectors?
   (d) Try to draw a circle that goes through all three vertices of the triangle.

Exercise 9   To Construct a Perpendicular to a Line from a Given Point Outside the Line (using a pair of compasses)

A 1. Let PQ be the line, and H, the point outside it.
   With centre H and radius any convenient size, draw arcs to cut the line at two points L and M.

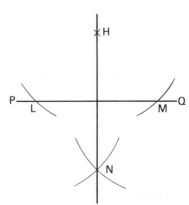

2. Now construct the perpendicular bisector of LM.

    (*a*) With centre L and radius bigger than $\frac{1}{2}$ LM, draw an arc below PQ.

    (*b*) Using the same radius but with centre M, draw another arc to cross the previous one (labelled N in the given diagram).

3. Join points H and N. This line is the required perpendicular bisector.

B Now repeat the construction of part A starting with a different line, AB, and a point T not on the line.

Exercise 10

Draw *free-hand sketches* from the given information. Make them to any suitable size. (If you draw a line to stand for a length of 4 cm then 8 cm lines should be roughly twice as long.)

1. Draw right-angled triangle ABC where AB = 8 cm, BC = 6 cm and AC = 10 cm. Draw the perpendicular bisectors of AB and AC. Let the bisectors meet at X.

2. Draw $\triangle$DEF where EF = 90 mm, ED = 70 mm and FD = 50 mm. Draw the perpendicular bisector of side EF and the bisector of angle DEF. Let the bisectors meet at Y.

3. Draw a line GH 9 cm in length. On GH, mark a point K where GK = 3 cm. Draw KL perpendicular to GH where KL = 2 cm.

4. Draw $\triangle$MNP where MN = 100 mm, MP = 40 mm and NP = 85 mm. Now draw a line PQ parallel to MN.

5. Draw $\triangle$RST where RS = 83 mm, RT = 35 mm and ST = 62 mm. Produce ST to a point U such that TU = 20 mm.

Exercise 11

Carry out these constructions but make free-hand sketches first:

1. Construct $\triangle$XYZ where YZ = 101 mm, XY = 80 mm and XZ = 83 mm. Draw a perpendicular from X to YZ using a set square and ruler. Let this perpendicular meet YZ at P. Measure XP.

2. Draw a straight line PQ = 82 mm. Mark a point T which is 49 mm from P and 56 mm from Q.

   Draw TU parallel to PQ using a set square and ruler. Let TU = 43 mm. Draw a perpendicular from T to PQ using a pair of compasses and let this perpendicular meet PQ at V. Join then measure UV.

3. Draw a straight line JR, 94 mm in length. On it, mark a point L where JL = 40 mm. At the point L, construct a perpendicular to JR, using a pair of compasses. On this perpendicular, mark a point M where LM = 67 mm. Measure:
   (a) JM   (b) RM

4. Without using a protractor, construct $\triangle$CDE where DE = 75 mm, $\angle$ CDE = 60° and CD = 40 mm. Measure CE.

5. Draw a straight line FG, 82 mm in length.
   Using a set square and ruler, draw a perpendicular to FG at F. Mark a point H on this perpendicular where FH = 40 mm. Join GH.
   Using a pair of compasses, construct $\angle$ GFI = 45° where I lies on GH. Measure FI.

10 Earning Money

Reminders: wage = hours worked × hourly rate

gross wage = basic wage + overtime

net wage = gross wage − deductions

net salary = gross salary − deductions

Exercise 1

1. Mrs Vale earned £5 an hour and worked for 7 h. How much did she earn altogether?

2. Mr Shaw earned £4.50 an hour and worked for 8 h. How much did he earn altogether?

3. How much did Vera earn if she worked for 6 h at £2.80 an hour?

4. Find Adam's earnings if he worked for 30 h at a rate of £4.50 an hour.

5. Find Mrs Dean's earnings if she worked for 28 h at £3.50 an hour.

6. How much does Mr Ellis earn in a week of 33 h working for £4.40 an hour?

7. Calculate Mr Heaton's earnings if he works 29 h for £5.35 an hour.

8. Mrs Upton works a 23-hour week. If she earns £4.79 per hour, calculate her total weekly earnings.

Exercise 2

1. If £36 is paid for 9 h work, how much is that per hour?

2. £45.50 is paid for 7 h work. How much is that per hour?

3. If £47.30 is paid for 10 h work, how much is that per hour?

4. If £216 is paid for 40 h work, how much is that per hour?

5. If £149.40 is paid for 36 h work, how much is that per hour?

Exercise 3

A Copy and complete the table:

	Gross wage	Deductions	Net wage
1.	£133	£38	
2.	£152.80	£46.90	
3.	£189.10		£131.65
4.		£51.25	£123.60

B **1.** If Mr Morgan's gross weekly wage is £129, find his net weekly earnings if deductions total £36.50.

2. Mrs York earns £651 per month gross. Calculate her net monthly earnings if deductions total £194.

3. Mr Scott's net weekly wage was £168.30 after deductions of £78.70. Find his gross weekly wage.

4. Mr Nugent's net salary was £4150 p.a. If deductions totalled £1570, find his gross salary.

5. Mrs Bell's basic wage was £96 per week. If she earned £19.20 overtime, find her gross wage.

6. Mrs Queen earned £138.50 per week gross. If this included £13.85 overtime, find her basic wage.

7. Mrs Jopson's gross salary was £7912. If deductions came to £2516, find her net salary.

8. Mr Parry's gross earnings were £606.67 per month. If his net earnings were £423.80 per month, calculate the deductions.

121

1. Mr Ayling's basic wage was £98 per week. If he then earned £28 overtime and deductions came to £33, calculate his net wage.

2. Mrs Boyd's basic wage was £129 per week. If she then earned £12.90 overtime and deductions came to £43.40, calculate her net wage.

3. Mr Chandler took home £113.70 after deductions of £51.90. If his gross pay included £16.56 overtime, calculate his basic pay.

4. Mr Donnelly's basic weekly wage was £148.80. If overtime came to £18.60 and his take-home pay was £117.90, calculate the deductions.

5. Mrs Emerson took home £112.85 after deductions of £55.90. If her basic wage was £135, calculate how much overtime she had earned.

Exercise 5 **M**

A Copy and complete the table:

	Name	Time worked (h) Mon	Tue	Wed	Thu	Fri	Total (h)	Hourly rate	Wage
e.g.	E. Gill	7	9	6	$7\frac{1}{2}$	8	$37\frac{1}{2}$	£3.60	£135
1.	A. Hunt	8	7	8	8	8		£4.30	
2.	B. Inman	7	7	7	8	7		£6.25	
3.	C. Jukes	7	$7\frac{1}{2}$	$7\frac{1}{2}$	8	8		£5.15	
4.	D. Kemp	$6\frac{1}{2}$	$7\frac{1}{2}$	7	7	8		£4.65	
5.	F. Law	$7\frac{1}{2}$	8	7	$6\frac{1}{2}$	$7\frac{1}{2}$		£5.40	

B 1. The table shows the length of time worked each day by a hairdresser:

(a) Calculate the total time worked in the week.

(b) Calculate her earnings at a rate of £5.50 an hour.

Day	Time
Mon	7 h 30 min
Tue	8 h 20 min
Wed	7 h 50 min
Thur	8 h 45 min
Fri	7 h 15 min
Sat	4 h 20 min

2. Here are the times Mr Foster worked each day for a week:

8 h 15 min	8 h 20 min	7 h 50 min
7 h 45 min	7 h 40 min	3 h 40 min

He did not work on Sunday.

(a) Calculate the total time he worked in the week.

(b) Calculate his earnings at a rate of £4.80 an hour.

Exercise 6

Copy and complete the time card below:

			F. G. Martin				
Day	Time in	Time out	Time worked h min	Time in	Time out	Time worked h min	Total daily time h min
Mon	07.00	12.00	5 00	14.00	16.00	2 00	7 00
Tue	08.00	12.00		13.30	17.00		
Wed	07.30	12.15		13.30	16.00		
Thu	08.50	11.50		13.00	17.30		
Fri	09.00	11.45		13.00	17.15		
Sat	08.00	12.00		—	—		
					Total weekly time =		

Exercise 7

A Calculate the overtime earnings:

1. 3 h overtime (double time) when the basic rate is £3.50 an hour,

2. 6 h overtime (time and a half) when the basic rate is £4 an hour,

3. 4 h overtime (double time) when the basic rate is £4.20 an hour,

4. 10 h overtime (double time) when the basic rate is £6.30 an hour,

5. 7 h overtime (time and a half) when the basic rate is £5.40 an hour.

B Copy and complete:

	Normal hours	Total hours worked	Basic hourly rate	Type of overtime	Gross weekly earnings
1.	40 h	45 h	£3.50	Double time	
2.	35 h	42 h	£5.10	Double time	
3.	37 h	39 h	£4.80	Time and a half	
4.	28 h	36 h	£3.90	Time and a half	
5.	38 h	50 h	£6.50	Time and a half	

Exercise 8

1. Patrick Nolan earns £129 for a basic working week of 30 h. During one week he works for 38 h and is paid overtime at double-time rate. Calculate:
   (a) Patrick's basic hourly rate,
   (b) the amount of overtime earned,
   (c) his gross weekly earnings.

2. Mrs Sagar's basic weekly wage is £144 for a 40-hour week. If she works 46 h during a certain week, being paid overtime at time and a half, calculate:
   (a) her basic hourly rate,
   (b) her gross weekly earnings.

3. Mr Owen's basic weekly wage is £179.20 for a 32-hour week. During a certain week he works for 39 h and is paid at double-time rate. Calculate his gross weekly earnings.

4. Mrs Rimmer earns £106.40 for a basic working week of 28 h. During a certain week she works for 32 h and is paid overtime at time and a half. Calculate her gross weekly earnings.

Exercise 9

A 1. How much per annum is:
  (a) £398 per month?
  (b) £509.50 per month?
  (c) £846.75 per month?
  (d) £107 per week?
  (e) £216 per week?
  (f) £147.50 per week?

B 1. Mr Milne earns £6552 per annum.
  (a) How much is that per month?
  (b) How much is that per week?

2. Mr Robertson earns £8580 per annum.
  (a) How much is that per month?
  (b) How much is that per week?

3. How much per month is £5739 p.a.?

4. How much per week is £4914 p.a.?

5. How much per week is £8148.40 p.a.?

C Who, in each of the following, is better paid?

1. Barry earns £5325 p.a. and Gavin earns £443.25 per month.

2. Avril earns £87.48 per week and Lucy earns £4550 p.a.

3. Mandy earns £352.75 per month and Mushtaq earns £4236 p.a.

4. Mr Waddell earns £169 per week and Mrs Thorpe earns £732 per month.

5. Mr Usher earns £144.25 per week and Mrs O'Brien earns £625.10 per month.

Commission, Piecework and Bonus

Commission is usually paid to salespeople. It is a percentage of the value of the goods they sell. They are normally paid a basic weekly wage or a salary; the commission is extra money to encourage them to sell more.

People who sell from mail order catalogues are paid commission. The commission in this case is often stated as 'so much in the pound'. This is the same as a percentage.

10 p in the pound is the same as 10%.
20 p in the pound is the same as 20%.

Changing percentages to so much in the pound may help in finding percentages of sums of money without a calculator, and can be particularly useful in estimating such percentages of sums of money.

e.g. 1 To estimate 15% of £6.95

$$15\% \text{ of } £6.95 \approx 7 \times 15p$$
$$= 105 p$$

so 15% of £6.95 is
<u>slightly less than £1.05</u>

> *Reasoning*
> £6.95 is almost £7
> 15% = 15 p in the pound
> that is, 15 p for £1
> which is $7 \times 15p$ for £7

e.g. 2 To estimate $22\frac{1}{2}\%$ of £38.75

$$22\tfrac{1}{2}\% \text{ of } £38.75 \approx 20\% \text{ of } £40$$
$$= 40 \times 20 p$$
$$= £8$$

so $22\frac{1}{2}\%$ of £38.75 is <u>about £8</u>

> Sometimes it is
> useful to round
> the percentage as
> well as the amount
> of money.

Exercise 10

Estimate

1. 30% of £6
2. 20% of £9
3. 40% of £19.80
4. 70% of £42.30
5. 30% of £12.99

6. 18% of £81.50
7. 11% of £139.70
8. 6% of £168.80
9. 29% of £74.68
10. 38% of £59.34

Exercise 11

Show how to key in the following percentages on a calculator:

A *e.g.* $46\% = \boxed{\cdot}\,\boxed{4}\,\boxed{6}$

(Although we normally write 46% as 0.46 rather than as .46 there is probably no need to key in the zero. Calculators normally insert the zero automatically.)

1. 82%   **3.** 74%   **5.** 37%   **7.** 70%

2. 91%   **4.** 11%   **6.** 24%   **8.** 99%

B *e.g. 1* $6\% = \boxed{\cdot}\,\boxed{0}\,\boxed{6}$ *e.g. 3* $17\frac{1}{2}\% = \boxed{\cdot}\,\boxed{1}\,\boxed{7}\,\boxed{5}$

e.g. 2 $1\% = \boxed{\cdot}\,\boxed{0}\,\boxed{1}$ *e.g. 4* $\frac{3}{4}\% = \boxed{\cdot}\,\boxed{0}\,\boxed{0}\,\boxed{7}\,\boxed{5}$

(*Note* $\frac{3}{4} = 0.75$ so $\frac{3}{4}\% = 0.75\%$.)

1. 8%   **4.** $22\frac{1}{2}\%$ **7.** $83\frac{1}{2}\%$ **10.** $\frac{1}{2}\%$

2. 2%   **5.** $37\frac{1}{2}\%$ **8.** $2\frac{1}{2}\%$ **11.** $\frac{1}{4}\%$

3. $12\frac{1}{2}\%$ **6.** $45\frac{1}{2}\%$ **9.** $7\frac{1}{2}\%$ **12.** $33\frac{1}{3}\%$

Exercise 12

e.g. $17\frac{1}{2}\%$ of £30 = $0.175 \times £30 = \underline{£5.25}$

(The above can be worked out on a calculator by keying in:
$\boxed{\cdot}\,\boxed{1}\,\boxed{7}\,\boxed{5}\,\boxed{\times}\,\boxed{3}\,\boxed{0}\,\boxed{=}$.
Note that some calculators have a percentage key*, $\boxed{\%}$.)

1. 65% of £38   **9.** 44% of £185.50

2. 82% of £47   **10.** 18% of £39.70

3. 39% of £71   **11.** 27% of £129.95

4. 35% of £26.80   **12.** 4% of £254.99

5. $27\frac{1}{2}\%$ of £56   **13.** $2\frac{1}{2}\%$ of £95.80

6. $12\frac{1}{2}\%$ of £14   **14.** $33\frac{1}{3}\%$ of £48

7. $54\frac{1}{2}\%$ of £94   **15.** $33\frac{1}{3}\%$ of £92.61

8. $7\frac{1}{2}\%$ of £86   **16.** $\frac{1}{2}\%$ of £450

*See Appendix 2, p.473.

Exercise 13

A Find the commission at 3% on goods worth:

   1. £100   **2.** £700   **3.** £250   **4.** £590   **5.** £467

B Calculate the commission, giving each amount correct to the nearest penny:

	Rate of commission	Value of goods			Rate of commission	Value of goods
1.	5%	£480		**6.**	4%	£459.90
2.	2%	£5160		**7.**	12%	£52.12
3.	7%	£156		**8.**	$7\frac{1}{2}\%$	£24
4.	10%	£92.60		**9.**	$2\frac{1}{2}\%$	£97
5.	15%	£65.20		**10.**	$\frac{1}{2}\%$	£499.95

C **1.** Mrs Cronin earns a basic wage of £80 and gets 2% commission. Calculate her earnings if she sells £3500 worth of goods.

   2. A salesman earns £45 per week plus commission of 4%. If he sells goods worth £2800, find his total earnings for the week.

   3. Mr Prescott earns £6900 p.a. and receives 5% commission on goods sold. Calculate his annual earnings if his sales total £25 000.

   4. Mrs Holt earns £65 per week and gets $2\frac{1}{2}\%$ commission. Calculate her earnings in a week when sales total £3642.

   5. Mr Iley is paid only commission at 2% and Mr Kerr is paid a basic wage of £69 plus commission at 1%. Who earns the most from selling goods worth £7290?

Piecework is a system of paying wages. There are many different kinds of piecework-rate systems. Each one is used to encourage the workers to work harder. In all the systems, the harder and faster a person works, the more items he or she produces and the more money he or she earns. In the basic piecework system, the piecework rates are usually given as 'so much per item', and we have:

> Earnings = no. of items produced × piecework rate

Other systems usually involve a standard length of time, normally called a standard minute (or SM). A particular job is split into small parts called elements. Each worker who carries out that job is then timed. For each person, the job is probably carried out for about an hour and an average time for each element calculated. An average is then found of all the workers' times for each element. By finding the total of the average times for each element, a total time for the whole job can be found. Allowances are normally added to this time, such as 'rest and fatigue factor' (that is, slowing down through tiredness) or 'machine delay' (that is, any factor beyond a worker's control). A time rating for the whole job can then be given, it is the time per unit. Since the timing is in standard minutes, the time for the task is given in SM per task. For example, in a factory 0.56 min may be allowed to hem two sleeves—so the time needed is stated as 0.56 SM per unit.

Each task is then given a payment rate in pence per SM. We then have:

> Earnings = no. of units produced × SM per unit × pence per SM

Exercise 14

A For each of the following, work out the earnings at the given piecework rates:

	Name	Piecework rate	Number of items completed
1.	K. Abbott	£3 each	34
2.	D. Brindle	24p each	400
3.	M. R. Dale	19p each	570
4.	G. Flynn	40 p per 100	28 000
5.	C. Forrest	35 p per 100	32 500
6.	D. W. Hanson	£2 per 1000	67 000
7.	F. Lovell	29 p per 10	3500
8.	C. Simpson	5.7 p each	1400

B 1. Morris works in a factory. His job is given a time rating of 0.5 SM per unit and he is paid 4 p per SM. Calculate his earnings when he completes:

(*a*) 800 units   (*b*) 1000 units   (*c*) 4200 units

2. Norman is paid 3.25 p per SM and does a job with a time rating of 1.5 SM per unit. How much does he earn for completing each of the following?

(*a*) 400 units   (*b*) 500 units   (*c*) 1800 units

3. Rita hems sleeves in a factory. In one week she completes 4000 units at 0.56 SM per unit and the rate of pay is 3.16 p per SM. Calculate her earnings.

4. Trimming, examining and folding overalls in a factory is given a time rating of 3.21 SM per unit. If Nimisha completes 850 units in one week and is paid 3.16 p per SM, calculate her earnings.

5. Denise topsews shirt collars in a factory. Calculate her earnings if she completes 5400 units and the time rating for her job is 0.49 SM per unit while the rate of pay is 3.16 p per SM.

A *bonus* is an extra payment given in addition to the basic wage. It is usually earned by meeting certain sales or production targets. Alternatively a bonus can be given for good attendance (for example, by not being absent from work for more than 1 h in a week).

Exercise 15

1. A person earning £90 in a week receives a bonus of 10%. Calculate:

(*a*) the bonus,   (*b*) the total weekly earnings.

2. A person earning £112 in a week receives a bonus of 10%. Calculate:

(*a*) the bonus,   (*b*) the total weekly earnings.

3. Julie earned £87.30 plus a 10% bonus. Find:

(*a*) her bonus,   (*b*) her total earnings.

4. Walter earned £147.50 plus a 12% bonus. Find:

(*a*) his bonus,   (*b*) his total earnings.

5. Zoe earned £98.50 plus an 8% bonus. Find:

    (*a*) her bonus,   (*b*) her total earnings.

6. The time rating for setting the zip on a zipped jacket was 0.78 SM per unit and the rate of pay was 3.28 p per SM. Kay completes 3140 units in a week and earns a 10% bonus. Calculate her total wages.

11 Spending Money

Household Budget

Exercise 1

A After deductions of income tax, National Insurance, etc., Ann had £69 take-home pay. This is how she spent it:

	£ p
Board and lodging	28.00
Fares and meals for a week	17.00
Clothing	6.00
Saving for holiday	5.00
Weekly expenses and entertainment	6.00
Emergency fund	3.00
Savings	?

1. How much does Ann save each week (not including holiday savings)?

2. Which three items of the seven items in the given list are essentials?

3. What do the extras (the non-essentials) total?

4. Write the total cost of the essentials as a fraction of the take-home pay, giving the fraction in its simplest terms.

B Assume that your take-home pay is £69 and that your fares and meals for a week only total £12.50. Make out a plan of your own using the same items as listed in part A. Decide for yourself how much to spend on each item. Be fair when estimating the board and lodging. (Find out from home the cost of food for one person, or work out the costs for yourself by planning a week's meals. Don't forget to cost drinks! Be sure to include the cost of the lodgings. Find out how much it costs to rent a flat or how much you'd pay towards costs involved in running the house in which you live. These costs include rent or mortgage, rates, heating and cooking.)

C Work out the total cost of taking sandwiches to work for 5 lunches. Decide what you would eat and cost the items (do not forget the bread and butter).

D Suppose you use a 200 g jar of coffee every four weeks, find the price of coffee in your local shops and work out your savings in one year if you bought the cheapest brand instead of the most expensive.

E Repeat part D if 200 g of coffee only lasts 26 days.

Some bills, such as gas, electricity and telephone bills, are normally paid quarterly (that is, four times a year). Some bills may be paid twice yearly, some monthly and some weekly. It is very worthwhile to plan ahead so that when a bill is received you already have the money to be able to pay it.

Exercise 2

The weekly amounts of money that need to be saved to pay bills, such as for gas and electricity, can be estimated by totalling the previous year's bills and dividing the total by 52. (If the money is to be saved monthly, the total should be divided by 12.)

Answer the following. Where necessary, give answers correct to the nearest penny.

1. Mr and Mrs West's gas bills for the year were £72.60, £70.20, £46.50 and £52.50.
   (a) Find the total cost of gas for the year.
   (b) If they saved monthly to pay the bills, how much did they need to save a month?
   (c) How much should they save each week to pay the bills?

2. Mr and Mrs Glenn's electricity bills for last year came to £71.21, £70.12, £52.62 and £69.93.
   (a) Find the total cost of electricity for the year.
   (b) If they saved monthly to pay the bills, how much did they need to save a month?
   (c) How much would they have needed to save each week to pay the bills?

3. Mr and Mrs Ewart's water services charge for a year was £106.20.
   (a) If the amount was paid in two equal instalments, how much would each instalment be?
   (b) If they saved monthly to pay the £106.20, how much should they have saved a month?
   (c) How much would they have needed to save each week to pay the full water services charge?

4. If Mr and Mrs Graham's contents insurance was £62.40 a year, find how much they needed to save to pay this amount if they saved the same amount:
   (a) each month,   (b) each week.

5. Mr and Mrs Imrie needed to pay £41.60 house insurance for the year, find how much they needed to save:
   (a) each month,   (b) each week.

6. Mr and Mrs Lamb's gas bills for a year were £201.74, £111.37, £43.90 and £125.50. To pay these bills, how much would they have needed to save:
   (a) each month,   (b) each week?

Exercise 3

A A copy of a family's household accounts for two weeks is given below:

Date	Income		£	p	Date	Expenditure		£	p
9 Oct	Balance	b/f	4	90	9 Oct	Essential savings		58	77
9 Oct	Wages		139	59	9 Oct	Fares and meals (work)		15	00
13 Oct	Child benefit		14	00	10 Oct	Food		44	70
					10 Oct	Other household goods		1	55
					10 Oct	Toiletries		2	45
					15 Oct	Medicine		2	80
					15 Oct	Newspapers		1	78
					15 Oct	Savings for bank		25	00
					16 Oct	Balance	c/f	6	44
	Total		158	49		*Total*		?	

134

Date	Income		£	p	Date	Expenditure		£	p
16 Oct	Balance	b/f	6	44	16 Oct	Essential savings		58	77
16 Oct	Wages		139	59	16 Oct	Fares and meals (work)		15	00
20 Oct	Child benefit		14	00	17 Oct	Food		45	36
					17 Oct	Other household goods		2	34
					17 Oct	Toiletries		1	90
					22 Oct	Newspapers		1	78
					22 Oct	Savings for bank		30	00
					23 Oct	Balance	c/f	?	
	Total		160	03		*Total*		160	03
23 Oct	Balance	b/f	?		23 Oct				

1. Total the expenditure column for the week beginning 9 October.

2. How much altogether was spent on 10 October?

3. The essential savings included the following: mortgage £19.45, rates £5.90, water rates £1.50, house insurance 80 p, contents insurance £1.20, life insurance £2.15, gas £6.25, electricity £4.50, TV rental £2.40, TV licence £1.12, together with a certain amount for clothes. How much was allowed for clothes?

4. What was the balance on 16 October?

B Make up a household account of your own. You will need to gather the information and make decisions as to how to spend the money. You may wish to cost the running of a car, hire purchase and other items in addition to some of those listed in part A.

Exercise 4 Best Buys

To save money it is worth working out the best buys. Do not be misled by the size of the packing. Recently, two different brands of washing powder were being sold in boxes that were almost identical in size. One box contained 4.65 kg of powder while the other contained 6.2 kg—a big difference! To find the best value for money it may be necessary to consider more than just the amount of contents and the cost. Other factors, such as the quality of the goods ought to be taken into consideration.

A Throughout this exercise, assume that the items being compared are of a similar quality. Find which is the better value for money.

1. 450 g for 53 p   or   225 g for 29 p
2. 750 g for £1.09   or   500 g for 94 p
3. 750 g for £1.09   or   600 g for 83 p
4. 250 mℓ for 69 p   or   200 mℓ for 55 p
5. 500 mℓ for 47 p   or   285 mℓ for 27 p
6. 500 mℓ for 56 p   or   1.25 ℓ for £1.39
7. 870 g for 75 p   or   930 g for 80 p
8. 4.65 kg for £3.89   or   6.2 kg for £4.89
9. 376 g for 41 p   or   220 g for 29 p
10. 510 g for 66 p   or   3 kg for £3.59

B While shopping, compare the prices of different-sized packs of goods and calculate which is the best value for money with regard to price, not quality.

Exercise 5   Holidays

A The average daily maximum temperatures in Calella (in Spain) and in London are given in the graph:

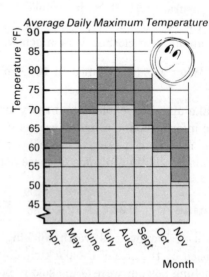

Average Daily Maximum Temperature

Key: ▢ = London

      ▢ = Calella

1. (*a*) Which months are the hottest in Calella?
   (*b*) Give the temperature in Calella during the hottest months.

136

2. What is the lowest temperature in each of the following?
(a) Calella
(b) London

3. (a) Find the temperature in London in June.
(b) Find the temperature in Calella in September.
(c) Use the formula

$$C = \tfrac{5}{9}(F - 32)$$

to change the temperatures of parts (a) and (b) into degrees Celsius.

B The following table gives the prices of some 14-day holidays in or near Calella:

14 days   Prices per person (pounds)

Departure Dates		Camping Bell Sol	2 bedded flatlets Bell Sol	2/3/4/5/6 bedded Holiday Centre apts	Hotel Balmes full board	Hotel Victoria full board
26 Apr, 3 May	30 Apr, 17 May	77	99	109	139	139
10 May	14 May	84	108	134	159	159
17, 24 May	21, 28 May	99	128	159	169	169
31 May, 7 June	4, 11 June	99	128	159	169	169
14, 21 June	18, 25 June	109	138	169	179	179
28 June, 5 July	2, 9 July	123	153	189	196	196
12 July	16 July	128	161	194	206	206
19, 26 July, 2 Aug	23, 30 July, 6 Aug	135	164	199	209	209
9, 16 Aug	13, 20 Aug	135	164	199	209	209
23 Aug	27 Aug	125	158	195	198	198
30 Aug	3 Sept	125	154	189	189	189
6 Sept	10 Sept	119	147	179	179	179
13, 20 Sept	17, 24 Sept	109	137	174	179	179
27 Sept, 4 Oct	1, 8 Oct	85	115	144	169	169

SUPPLEMENTS* (Prices per person per day):
Luxury camping: 3 sharing 50 p, 2 sharing £1.25.
2/3 Holiday Centre studios: 3 sharing NIL, 2 sharing £1.25.
4/5 Holiday Centre apartments: 3 sharing £2.00.

*A supplement is an additional charge.

Use the table on the previous page to answer the following:

1. What is the cost per person for camping if there are no supplements to pay and if the departure date is 5 July?

2. What is the cost for someone to depart on 21 May and to stay at the Hotel Balmes?

3. What is the cost for someone to depart during the first week in August and stay at the Hotel Victoria?

4. What is the difference in costs between the Hotel Victoria and Camping Bell Sol if no supplements are payable and if the departure date is 16 July?

5. What is the cost per person to depart on 23 August and to stay at a 4/5 Holiday Centre apartment if 3 people are sharing? (Do not forget the supplement. This extra cost must be worked out for a 14-day holiday.)

6. (a) What is the total cost for 2 people to depart on 6 August and to stay in a 2/3 bedded Holiday Centre studio?
   (b) How much would 2 people in part (a) have saved if they stayed at a 2-bedded flatlet instead of at a studio?

7. What is the total cost for Mr and Mrs Houghton and their two daughters to depart on 23 July and stay at the Hotel Balmes if there are no discounts and no supplements?

8. What is the total cost for Mr and Mrs Ramsay and their 9-year-old son Michael to depart on 11 June and to stay at the Hotel Victoria, if children under 12 years of age get a 15% discount?

Foreign Currencies

When travelling abroad you need to change your money (pounds sterling, £) into foreign currency. Banks will do this for you. They have a list of *rates of exchange*. These rates can change from day to day. A selection of rates of exchange is given on the next page.

Rates of Exchange Equivalent to £1 Sterling*

Country	Currency	Country	Currency
Australia	2.07 dollars	Ireland	1.15 punts
Austria	24.65 schillings	Italy	2410.00 lire
Belgium	72.30 francs	Malta	0.64 pounds
Canada	2.01 dollars	New Zealand	2.77 dollars
Denmark	12.95 kroner	Norway	10.82 kroner
Finland	7.78 markkaa	Portugal	221.00 escudos
France	10.75 francs	Spain	219.00 pesetas
W Germany	3.50 Deutschmarks	Sweden	10.90 kronor
Greece	250.00 drachmas	Switzerland	2.95 francs
Holland	3.94 guilder	USA	1.4325 dollars
Iceland	60.00 krónur	Yugoslavia	515 dinars

Amount of foreign currency = number of pounds × rate of exchange

Exercise 6

Answer the following using the rates of exchange given:

1. Change £10 into:
   - (a) Australian dollars,
   - (b) guilder
   - (c) Belgian francs,
   - (d) dinars

2. Change £30 into:
   - (a) schillings,
   - (b) Maltese pounds,
   - (c) punts,
   - (d) drachmas.

3. Change £35 into:
   - (a) Swiss francs,
   - (b) Norwegian kroner,
   - (c) Deutschmarks,
   - (d) US dollars.

4. Mrs Clark changes £50 into francs before crossing the Channel to France. How many francs did she get?

*See the glossary, p.480.

5. Mr Barnes changed £85 into kronor before going to Sweden. How many did he get?

6. A family changed £19 into escudos, how many did they get?

7. Anna changed £120 into lire while in Italy. How many lire did she get?

8. A family took £97 in pesetas to Spain when going on holiday. How many pesetas was that?

While on holiday in a foreign country you would probably see prices and think, 'What is that in pounds?'

To find a cost in pounds sterling, use the formula:

$$\text{Number of pounds} = \frac{\text{amount of foreign exchange}}{\text{rate of exchange per pound}}$$

Exercise 7

1. Change into pounds sterling, giving answers to the nearest penny:
   (a) 420 krónur,
   (b) 80 Canadian dollars,
   (c) 300 Danish kroner,
   (d) 180 New Zealand dollars,
   (e) 389 markkaa,
   (f) 125 Swiss francs,
   (g) 50 000 pesetas,
   (h) 1460 schillings.

2. Some governments limit the amount of their currency you can take into their country.
   The present rates per person are given. Change each amount into pounds sterling.
   (a) Greece, 3000 drachmas,
   (b) Yugoslavia, 2500 dinars,
   (c) Portugal, 5000 escudos,
   (d) Malta, 50 pounds,
   (e) Iceland, 3000 krónur,
   (f) Italy, 400 000 lire,
   (g) Spain, 150 000 pesetas.

3. Anthony spent 75 guilder while in Holland. How much is that in pounds sterling?

4. On holiday, Carol bought a present costing 21 Deutschmarks. How much did it cost in pounds sterling?

5. Mr Ward bought a watch in Switzerland for 46.40 francs. What was its cost in pounds sterling?

6. A sweater cost 420 kroner in Norway. How much did it cost in pounds sterling?

7. A carved wooden bowl cost 2480 dinars in Yugoslavia. How much is that in pounds sterling?

8. A crystal decanter cost 350 kronor in Sweden. How many pounds sterling is that?

9. A lace cloth from Belgium cost 5268 Belgian francs. Calculate its cost in pounds sterling.

10. A leather bag cost 85 300 lire in Italy, while a similar sort of bag cost 7980 pesetas in Spain. Which was the cheaper buy and by how many pounds sterling?

Although exchange rates are normally given as an amount that is equivalent to £1 sterling, occasionally, the rate may be quoted differently. If so, then the previous formulae will not work.

Exercise 8

Answer the following using the exchange rates given in each question:

1. Find the cost, to the nearest penny, of an item costing 8 Norwegian kroner if the exchange rate is 9.2 p to 1 kroner.

2. An article costs 5.30 markkaa. Find its cost in pence if the exchange rate is 12.9 p to 1 markkaa.

3. At an exchange rate of 69.8 p to 1 US dollar, find the cost in pounds sterling of an item costing 4 dollars.

4. If the exchange rate is 4.1 p to 1 schilling, find the cost in pence of an article costing 9.40 schillings.

5. An article costs 2.85 guilder. Find its cost in pence using an exchange rate of 25.4 p to 1 guilder.

12 Loci

A door, labelled ABCD, is hinged
along BC.
If the door is opened, the locus of
the handle H is an arc of a circle.

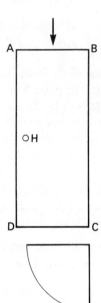

The locus of H is shown here:

Exercise 1

1. Sketch the locus of a mark on
the seat of a swing:

2. Sketch the locus of a mark on the edge of a see-saw:

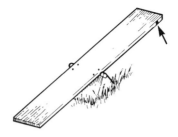

3. Sketch the locus of the handle on a sliding door when the door is opened.

Exercise 2

1. Draw the locus of the centre of a variable circle which always touches the same fixed line at the same fixed point ('variable' means that its size changes).
One possible position is shown in the diagram. P is the fixed point.

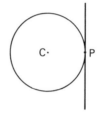

2. Draw the locus of the centre C of a circle of constant radius 25 mm that always passes through a fixed point P.

3. A and B are two fixed points that are 5 cm apart.
Draw the locus of the centre C of a variable circle that always passes through the two fixed points A and B.

4. Draw the locus of the centre C of a circle with constant radius 20 mm which always touches externally a fixed circle with radius 15 mm.

5. A groundsman was asked to mark out the shooting circle on a hockey pitch.
He was told to mark a line that was always 16 yd from the goal line. (The goal line measures 4 yd.)
Make a scale drawing to show the shooting circle. (Note that it is not actually a circle!)

The sketch shows five possible points on the shooting circle.

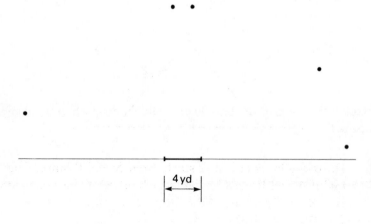

4 yd

Exercise 3

Throughout this exercise, you may find it useful to use geo-strips to help you to see the result.
If sizes are not given in a question then choose your own.

1. A metal rod is fixed at F.
Draw the locus of the other
end, B, as the rod is moved.

F●————————————————●B

2. PB is a metal rod which passes through a fitting F about which it can turn. The position of F remains fixed.

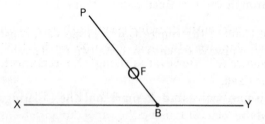

B slides along a horizontal bar, XY. Draw the locus of P. Make XY 12 cm long. Position F 1.5 cm directly above the centre of XY. PB should be 6 cm long.

3. (a) AP and BP are metal rods of equal length that are hinged at P. A and B are free to move along a fixed horizontal bar, XY.

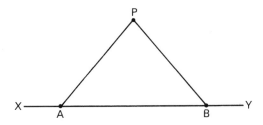

Draw the locus of P if A starts from X and B starts from Y and if A and B move towards each other at the same rate. Let XY = 8 cm and AP = BP = 4 cm.

(b) One type of car jack is shown in the diagram:

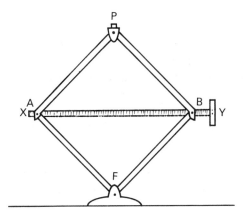

The top part of the drawing is similar to the rods in part (a). A and B are free to move towards each other along the horizontal XY. However, although XY remains horizontal, it is not fixed.

F is a fixed point and PA = PB = FA = FB.

Draw the locus of P using the same dimensions as in part (a).

4. Repeat question 3(a) using the same dimensions, but this time let A be fixed at position X. B should then start at Y and move towards A. Once again, draw the locus of P.

5. In book 3G, p.94, a diagram was given of a car jack that would not work properly.

The car jack shown in this diagram should work. Note that QP = QR = QS.

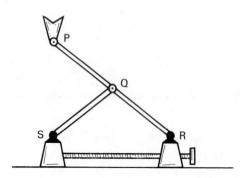

(*a*) Make a drawing to show the locus of P. S is fixed. R should be moved towards S. Make QP = QR = QS = 4 cm. RS can be any size up to 8 cm.

(*b*) Repeat part (*a*) but this time let QP = 3 cm, QR = 5 cm and QS = 4 cm.

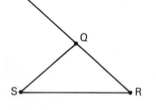

(*c*) Repeat part (*a*) but this time let QP = 5 cm, QR = 3 cm and QS = 4 cm.

Exercise 4 M

1. (*a*) Draw a pair of axes such that the numbers on both the *x*- and *y*-axes range from 0 to 7.

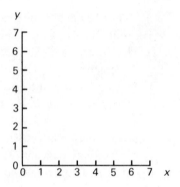

(b) Copy and complete the following table such that the x- and y-values always add up to 5 (i.e., $x + y = 5$).

x	0	1	2	3	4	5
y					1	

(c) Draw the locus of a point which moves so that the sum of the x- and y-co-ordinates is always 5.

(d) Using the same pair of axes, draw the locus of a point which moves so that it is always two units from the x-axis.

2. (a) Draw a pair of axes such that both the x- and the y-axis range from 0 to 40.

(b) Copy and complete the following table such that the product of the x- and y-values is always 12 (i.e., $xy = 12$).

x	1	2	3	4	6		24	36		$\frac{1}{2}$	40	30	20	0.6	0.4	0.3
y				3		1			36	24		0.4				

(c) Draw the locus of a point which moves so that the product of its x- and y-co-ordinates is always 12.

3. (a) Draw a pair of axes such that the x-axis ranges from $^-7$ to $^+7$ ($^-7 \leqslant x \leqslant 7$) and the y-axis ranges from $^-30$ to $^+20$ ($^-30 \leqslant x \leqslant 20$).

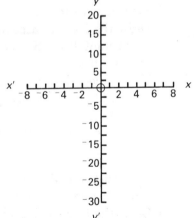

(b) Copy and complete the following table such that the y-value is 5 less than 3 times the x-value (i.e. $y = 3x - 5$):

x	$^-7$	$^-6$	$^-5$	$^-4$	$^-3$	$^-2$	$^-1$	0	1	2	3	4	5	6	7
$y = 3x - 5$		$^-23$										7			

(c) Draw the locus of a point which moves so that the *y*-co-ordinate is always 5 less than 3 times the *x*-co-ordinate.

Intersecting Loci

Exercise 5

1. (a) Copy △ABC.

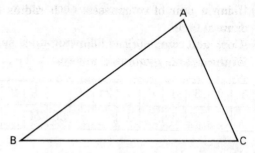

(b) Draw the locus of a point which moves so that it is always 3 cm from vertex A.

(c) Draw the locus of a point which moves so that it is always equidistant from B and C ('equidistant' means 'the same distance from').

(d) The two loci should intersect at two points X and Y where X lies inside the triangle. Mark points X and Y on your diagram.

(e) Write the length of BX.

(f) How long is BY?

2. (a) Construct △PQR where PQ = 55 mm, QP̂R = 60° and PR = 45 mm.

(b) Draw the locus of a point which moves so that it is always 35 mm from vertex R.

(c) Draw the locus of a point which moves so that it is always 20 mm from side PQ.

(d) Label the two points at which the loci from parts (b) and (c) intersect as X and Y.

(e) Write the lengths of QX and QY.

Exercise 6

1. (a) Draw any triangle and label it ABC.
    (b) Draw the locus of a point which moves so that it is equidistant from vertices A and B.
    (c) Draw the locus of a point which moves so that it is equidistant from vertices B and C.
    (d) Draw the locus of a point which moves so that it is equidistant from vertices A and C.
    (e) The three loci should meet at one point. If they do not, then check them. Label the point of intersection as X.
    (f) Using a pair of compasses, with radius AX and centre X, draw a circle.
    (g) Write what you notice about the circle and triangle.
    (h) What name is given to the circle?
    (i) What name is given to point X?

2. (a) Draw any triangle and label it DEF.
    (b) Draw the locus of a point which moves so that it is equidistant from sides DE and DF.
    (c) Draw the locus of a point which moves so that it is equidistant from sides ED and EF.
    (d) Draw the locus of a point which moves so that it is equidistant from sides FD and FE.
    (e) The three loci should meet at one point. If they do not, then check them. Label the point of interesection as P.
    (f) Using a pair of compasses and point P as centre, draw a circle that touches side DE.
    (g) Write what you notice about the circle and triangle.
    (h) What name is given to the circle?
    (i) What name is given to point P?

Exercise 7 M

1. Two trees are 7 m apart. If some treasure was buried 3 m from one tree and 6 m from the other, make a scale drawing and show four possible positions for the treasure.
    (Use a scale of 1 cm to 1 m.)

2. There is a ditch at the bottom of Mrs Kelly's garden. 5 m from the ditch is a tree.

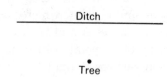

A landscape gardener was given the task of planting another tree. Mrs Kelly gave him these instructions:
'The new tree should be planted 3 m away from both the old tree and the ditch.' Make a scale drawing to show two possible positions for the new tree. (Use a scale of 1 cm to 1 m.)

3. Copy the map, but make all distances twice as big as those used in the diagram.
Note that the farmhouse is 30 m from the river and 60 m from the shop. Also, the shop is 50 m from the river.

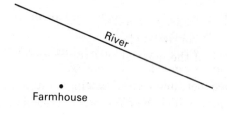

Tim went camping and pitched his tent so that it was equidistant from the farmhouse and the shop; it was also 20 m from the river. Make a scale drawing to show the position of his tent.

13 **Area and Volume**

Rectangular Areas

Exercise 1

1. A rectangle is shown here drawn on dotty paper:

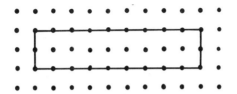

If the dots were spaced 1 cm apart, find:
(a) the area of the rectangle,
(b) the length of the perimeter of the rectangle.

2. Using dotty paper, draw the following:
(a) A rectangle which has the same area as the rectangle in question 1 but a different length of perimeter.
(b) A rectangle which has the same perimeter as the rectangle in question 1 but a different area.

3. (a) A square has a perimeter of 20 cm, find its area.
(b) A square has an area of 81 cm², find its perimeter.
(c) A square has an area of 625 cm², find its perimeter.

4. A rectangle has an area of 63 m². If it is 9 m long, find:
(a) its width,   (b) its perimeter.

5. The perimeter of a rectangle is 26 cm. It is 9 cm long.
(a) Find its width.
(b) Find its area.
(c) Find the length of side of a square that has the same area as the rectangle

6. Calculate the area of:
   (*a*) A rectangle with length = 6 cm and breadth = 5 cm,
   (*b*) A rectangle with length = 0.6 cm and breadth = 0.5 cm.

7. Calculate the area of a square with sides:
   (*a*) 3 m   (*b*) 0.3 m   (*c*) 0.03 m

8. (*a*) A rectangle has an area of 13.92 cm^2. If it is 2.9 cm wide, calculate its length.
   (*b*) A rectangle has an area of 324 m^2. If its length is 21.9 m, find its width correct to three significant figures.

Exercise 2

1. The house shown has four large windows at the front, each measuring 1.8 m by 1.2 m, and a small window measuring 1.2 m by 0.8 m. Find:
   (*a*) the area of one large window,
   (*b*) the area of the small window,
   (*c*) the total area of glass needed for the five windows at the front of the house.

2. A carpet measures 6 m by 4 m.
   (*a*) Calculate its area.
   (*b*) Calculate its cost if it is sold at £9.80 per square metre.
   (*c*) Calculate the cost of buying the carpet and having it fitted if the firm charges 80 p per square metre fitting charge.

3. A carpet measuring 5 m by 4 m is cut to fit a room 4.8 m long and 3.9 m wide. Calculate the area of carpet wasted.

4. If a piece of paper measuring 35 cm by 20 cm was cut into 5 cm squares, how many squares would there be?

5. If a piece of paper measuring 35 cm by 20 cm was cut into 2 cm squares:
   (*a*) How many squares would there be?
   (*b*) What area of paper would be left?

6. If a piece of paper measuring 35 cm by 20 cm was cut into 3 cm squares:
   (a) How many squares would there be?
   (b) What area of paper would be left?

7. A hockey pitch measures 100 m by 60 m.
   (a) Calculate its area.
   (b) If, when a groundsman mows the grass, each strip cut by the mower is 40 cm wide, how far will he walk in mowing the whole hockey pitch?

8. (a) Calculate the area of the room shown here:
   (b) If carpet costs £9.50 per square metre and under-lay a further £3 per square metre, calculate the total cost of carpet and underlay for this room.

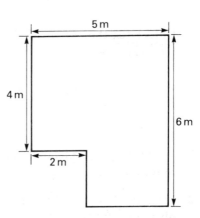

Exercise 3   Areas of Parallelograms, Triangles and Trapezia

A 1. Calculate the following for the given parallelogram:
   (a) the area,
   (b) the perimeter.

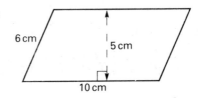

2. Find the area of:
   (a) a parallelogram with base = 7 m and perpendicular height = 6 m,
   (b) a parallelogram with base = 6 cm and perpendicular height = 4.7 cm,
   (c) a parallelogram with base = 3.4 m and altitude = 1.9 m.

3. For the given parallelogram, find:

(a) its area,

(b) its perimeter.

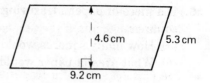

4. The area of a parallelogram is $63\,m^2$. If its base is $9\,m$, find its height.

5. A parallelogram has an area of $12.04\,m^2$. If its base is $4.3\,m$, find its height.

B **1.** Calculate the area of the triangle given:

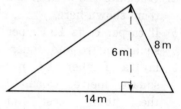

2. Find the area of a triangle whose base is $16\,cm$ and perpendicular height $12\,cm$.

3. Find the area of a triangle whose base is $68\,mm$ and whose altitude is $26\,mm$.

4. A triangle has an area of $24\,cm^2$. If its base measures $8\,cm$, find its altitude.

5. A triangle has an area of $20.54\,m^2$. Find the length of its base if its perpendicular height is $5.2\,m$.

C **1.** Calculate the area of each of the following trapezia:

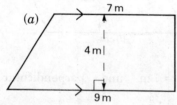

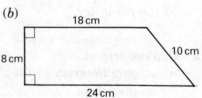

2. The parallel sides of a trapezium are $14\,cm$ and $6\,cm$. If the perpendicular distance between the parallel sides is $5\,cm$, find the area of the trapezium.

3. Find the area of a trapezium with parallel sides measuring 46 mm and 68 mm if the perpendicular distance between them is 50 mm.

4. The parallel sides of a trapezium measure 7.8 cm and 5.2 cm. If the perpendicular distance between the parallel sides is 3.4 cm, calculate the area of the trapezium.

Exercise 4   Areas of Borders

A The diagram shows a square lawn with side y metres surrounded by a path of constant width. The path and lawn together form a square with side x metres. Answer the following for a garden; $x = 12$, $y = 8$ and the lengths are in metres.

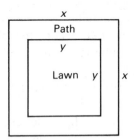

1. (a) What is the area of the lawn?
   (b) What is the total area of the lawn and path?
   (c) Using the answers to parts (a) and (b), find the area of the path.

2. The area of the lawn and path $= x^2$ square metres
   and the area of the lawn $= y^2$ square metres
   so the area of the path $= x^2 - y^2$ square metres
   Work out the area of the path using the fact that
   $$x^2 - y^2 = (x + y)(x - y).$$

3. Here is another drawing of the lawn and path. This time, the path has been divided into four identical trapezia.
   (a) Use the formula,
   $$A = \tfrac{1}{2}(x + y)h$$
   to find the area of one trapezium.
   (b) Use the answer to part (a) to find the area of the path.

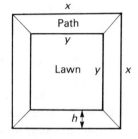

4. The four trapezia can be re-positioned to make a parallelogram:

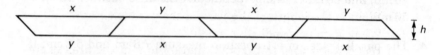

(a) What is the length of the base of this parallelogram?

(b) What is its perpendicular height?

(c) Use the answers to parts (a) and (b) to find the area of the path.

5. In this drawing of the lawn and path a line has been marked following the centre of the path. The line shows the average length of the path.

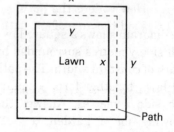

(a) Find the average length of the path.

(b) How wide is the path?

(c) Using the answers to parts (a) and (b), find the area of the path.

B The diagram shows a lawn measuring 8 m by 6 m surrounded by a path that is 2 m wide.

1. (a) What is the area of the lawn?

(b) What is the area of the lawn and path?

(c) Use the answers to parts (a) and (b) to find the area of the path.

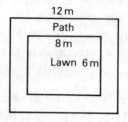

2. (a) Make a sketch of the lawn and path and divide the path into four trapezia (two large and two small).

(b) What is the area of one of the large trapezia?

(c) What is the area of one of the small trapezia?

(d) Use the answers to parts (b) and (c) to find the area of the path.

156

3. (a) Make a sketch showing the four trapezia positioned to give a parallelogram (as in part A, question 4).

(b) What is the length of the base of the parallelogram?

(c) What is its perpendicular height?

(d) Use the answers to parts (b) and (c) to find the area of the path.

4. (a) Make a sketch of the lawn and path. Draw a broken line to show the average length of the path (as in part A, question 5).

(b) Find the average length of the path.

(c) How wide is the path?

(d) Find the area of the path using your answers to parts (b) and (c).

Circumference and Area of a Circle

Exercise 5

A Find the circumference and area of each of the following circles with dimensions as given. Use $\pi = 3.14$. Give answers to three significant figures.

1. Radius $= 2\,m$

2. Diameter $= 40\,cm$

3. Radius $= 34\,mm$

4. Radius $= 5.2\,cm$

5. Diameter $= 12.8\,m$

6. Radius $= 87\,mm$

7. Diameter $= 49.6\,cm$

8. Diameter $= 280\,mm$

B Find the circumference and area of each of these circles with dimensions as given. (Use $\pi = 3\frac{1}{7}$.)

1. Diameter $= 42\,cm$

2. Radius $= 35\,mm$

3. Radius $= 10\frac{1}{2}\,cm$

4. Diameter $= 56\,mm$

Exercise 6

1. A firm makes the 'rubbers' that fit around pram and push-chair wheels. What length of rubber is used to fit around a wheel having a diameter of:

(a) 25 cm?   (b) 18 cm?

(Use $\pi = 3.14$ and give answers to three significant figures.)

2. A lampshade has wire edging around its top and bottom edges. Using π = 3.142, find, to three significant figures, the length of wire needed for:

    (*a*) the top edge if its diameter is 20 cm,

    (*b*) the bottom edge if its radius is 14.5 cm.

3. The circular wire rim on a sieve has a diameter of 115 mm. Find the total length of wire needed to make rims for 50 sieves. Use π = 3.14 and give your answer to three significant figures.

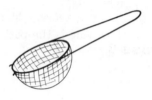

4. The handle of a paint tin is made from a thin metal strip that is bent to the shape of a semi-circle. Find the length needed to make 75 handles if the tins have a diameter of 150 mm. Use π = 3.14 and give your answer to three significant figures.

Exercise 7

Use π = 3.14 throughout this exercise and give answers to three significant figures:

1. A dart-board has a radius of 25 cm. Find its area.

2. Find the area of a circular rug with diameter 54 in.

3. A 'half-moon' rug has a diameter of 1.36 m. Find its area.

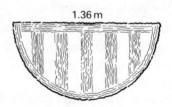

1.36 m

4. A wooden stool has a circular top. Find the area of the top if its diameter is 31 cm.

5. A hole is drilled through a circular metal plate. If the hole has a diameter of 0.6 cm and the plate a radius of 13 cm, calculate the area of the plate after the hole has been drilled.

Exercise 8

1. Will four circles, each of radius 2 cm, have the same total area as one circle of radius 8 cm?

2. Find the area of the largest circle that will fit inside a square of 112 cm perimeter. (Use $\pi = \frac{22}{7}$.)

3. Find the shaded areas. Use $\pi = 3.14$.

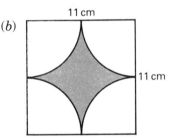

(a) 11 cm 11 cm

(b) 11 cm 11 cm

Exercise 9

A 1. An arc subtends an angle of 120° at the centre of a circle. What fraction of the circumference of the circle is the arc?

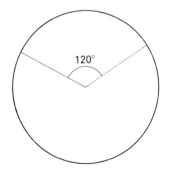

120°

2. A sector is formed by an arc and two radii. If the arc subtends an angle of 60° at the centre of the circle, what fraction of the circle is the sector?

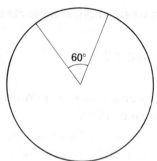

3. What fraction of a circle is each sector if its arc subtends the following angles at the centre of the circle?

(a) 90°   (b) 45°   (c) 150°   (d) 72°   (e) 240°

B For each question, using the value of π given, find:
(a) the length of the arc
(b) the area of the sector created by two radii that form the given angle at the centre of the circle.

1. Radius   = 6 cm   angle subtended = 90°   π = 3
2. Radius   = 20 cm   angle subtended = 36°   π = 3.14
3. Radius   = 8.2 m   angle subtended = 60°   π = 3
4. Radius   = 9.6 cm   angle subtended = 45°   π = 3.14
5. Diameter = 12 m   angle subtended = 150°   π = 3.14
6. Diameter = 2.4 m   angle subtended = 135°   π = 3.14
7. Radius   = 5.4 cm   angle subtended = 40°   π = 3.142
8. Diameter = 21.6 cm   angle subtended = 140°   π = $\frac{22}{7}$

Exercise 10

The circumferences of several circles are given. For each one, find the diameter. Where necessary, give your answer to three significant figures.

	Circumference	π
1.	12 cm	3
2.	576 mm	3
3.	25.8 m	3
4.	496 mm	3.14
5.	32 cm	3.14

	Circumference	π
6.	144 cm	3.14
7.	12 cm	3.142
8.	597 mm	3.142
9.	198 cm	$3\frac{1}{7}$
10.	308 mm	$\frac{22}{7}$

Volume and Surface Area of a Cuboid

A The cuboid shown here has a volume of 48 cm³:

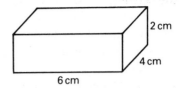

2 cm
4 cm
6 cm

1. Sketch three different cuboids having a volume of 48 cm³ and having the same base area as the cuboid shown. Write the dimensions on each sketch.

2. What do you notice about the heights of the cuboids in question 1?

B 1. Copy and complete the following table for the cuboids given:

Length l (cm)	Breadth b (cm)	Height h (cm)	Area of base $A=lb$ (cm²)	Volume $V=lbh$ (cm³)	Ah (cm³)
3	2	2			
4	4	3			
6	3	4			
10	5	3			
8	6	2.5			
7	3.5	2			
9	4.8	3.1			
4.7	3.4	2.9			

2. Write what you notice about the values in the last two columns.

Volume of a cuboid $V = lbh$

so $V = Ah$ (since $A = lb$)

Volume of a cuboid $V = lbh = Ah$

Exercise 12

1. Here is one possible net of a cube:
   If each edge of the cube measured 5 cm, find:
   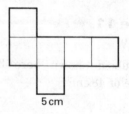
   (a) the area of one face,
   (b) the total surface area of the cube,
   (c) the volume of the cube.

2. If each edge of a cube measured 2.5 cm, find:
   (a) the area of one face,
   (b) the total surface area of the cube,
   (c) the volume of the cube.

3. If each edge of a cube measured x cm, find:
   (a) the area of one face,
   (b) the total surface area of the cube,
   (c) the volume of the cube.

4. If each edge of a cube measured $2x$ cm, find:
   (a) the area of one face,
   (b) the total surface area of the cube,
   (c) the volume of the cube.

5. A cube has a total surface area of 54 cm^2.
   (a) What is the area of one face?
   (b) What is the volume of the cube?

6. A cube has a volume of 8 cm^3.
   (a) What is the length of each edge?
   (b) What is the area of one face?
   (c) What is the total surface area of the cube?

7. If the given net was cut out of a rectangular piece of card to make a cube with an edge of 1 cm, the card would have an area of 12 cm^2.
   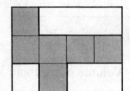
   Find a different net of a cube that can be cut out of a rectangular piece of card of 10 cm^2 area to make a cube with 1 cm edges.

8. What are the dimensions of the smallest rectangular piece of card that can be used to make two cubes each having 1 cm edges?

9. What are the dimensions of the smallest rectangular piece of card that can be used to make two, open cube-shaped boxes each having 1 cm edges?

10. The sketch shows the net of a cuboid:

Calculate:
(*a*) its total surface area,
(*b*) the volume of the cuboid.

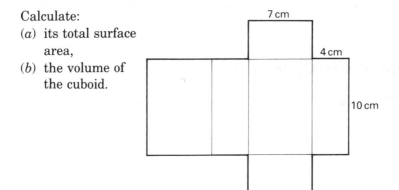

7 cm
4 cm
10 cm

Exercise 13

1. A tinned steel baking tin measures 30 cm × 30 cm × 3 cm. Calculate the area of the tinned steel that was used to make it.

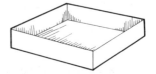

2. The internal measurements of a roasting tin are 35 cm by 28 cm by 6 cm deep. The inner surface is given a non-stick coating. Calculate the area coated.

3. A metal bar in the shape of a cuboid is 80 cm long, 20 cm wide and 10 cm thick.
(*a*) Calculate its volume.
(*b*) If the density of the metal is 7000 kg/m³ (i.e., every cubic metre weighs 7000 kg or every cubic centimetre weighs 7 g), work out how heavy the metal bar is giving the answer in kilograms.

4. A plastic container is 20 cm long, 15 cm wide and 5 cm deep.

(a) Calculate its volume.

(b) If the container is full of water, how many litres of water are in it?

(*Note* 1000 cm³ holds 1 ℓ.)

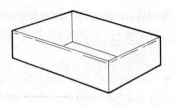

5. A concrete paving stone measures 600 mm by 600 mm by 50 mm. How heavy is it if concrete has a density of 2016 kg/m³ (i.e., 1 cm³ has a mass of 2.016 g)?

Volume of a Cylinder

The rectangle and circle each have an area of 4 cm².

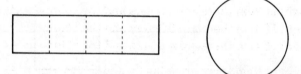

When a 1 cm-layer is built on to them, their volume = 4 cm³.

Volume of cuboid = 4 cm³   Volume of cylinder = 4 cm³

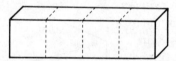 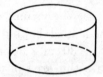

For areas of 6 cm² a 1 cm-layer has a volume of 6 cm³.
For areas of 15 cm² a 1 cm-layer has a volume of 15 cm³.
For areas of A cm² a 1 cm-layer has a volume of A cm³.

If they are made into taller prisms with another 1 cm-layer, then their volume is doubled:

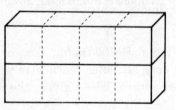 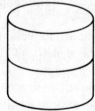

If the volume of 1 layer $= A\,\text{cm}^3$

the volume of 2 layers $= A \times 2\,\text{cm}^3$

the volume of 3 layers $= A \times 3\,\text{cm}^3$

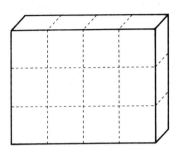 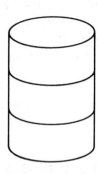

Summing up:

For a height of 1 cm there is 1 layer and the volume $= A\,\text{cm}^3$.
For a height of 2 cm there are 2 layers and the volume $= A \times 2\,\text{cm}^3$.
For a height of 3 cm there are 3 layers and the volume $= A \times 3\,\text{cm}^3$.
For a height of 4 cm there are 4 layers and the volume $= A \times 4\,\text{cm}^3$.

For a height of h cm there are h layers and the volume $= A \times h\,\text{cm}^3$.

So if the area of the base $= A$ and the height $= h$

the volume of a cuboid $= Ah$

and also the volume of a cylinder $= Ah$

The same reasoning may be applied to any prism, so we have, volume of a prism $= Ah$.

> Volume of a cylinder $V = Ah$
>
> where A is the area of the base circle and h is the perpendicular height.

Since the base of a cylinder is a circle and since the area of a circle $A = \pi r^2$, the volume of a cylinder can also be found using the formula:

> Volume of a cylinder $V = \pi r^2 h$
>
> where r is the radius of the base circle and h is the perpendicular height.

The cross-section* of a cylinder is a circle having the same area as the base of the cylinder, so the formula for the volume of a cylinder can also be written as:

$$\text{Volume of a cylinder} = \text{area of cross-section} \times \text{perpendicular height.}$$

Exercise 14

A Calculate the volume of each of the following cylinders. Use the value of π given and where necessary, round answers to three significant figures.

	Radius	Height	π
1.	2 cm	5 cm	3
2.	3 m	3 m	3
3.	3 m	7 m	3
4.	7 cm	4 cm	3
5.	10 cm	3.5 cm	3
6.	1 m	2 m	3.14
7.	4 cm	5 cm	3.14
8.	5 cm	6 cm	3.14

	Radius	Height	π
9.	10 cm	4 cm	3.14
10.	6 cm	4.3 cm	3.14
11.	1.4 m	1.2 m	3.14
12.	3.8 cm	5.7 cm	3.14
13.	9 cm	3 cm	3.142
14.	0.8 m	0.6 m	3.142
15.	7 cm	5 cm	$\frac{22}{7}$
16.	$2\frac{1}{2}$ cm	7 cm	$3\frac{1}{7}$

B Answer these. Give answers to three significant figures. If a diameter is given then do not forget to find the radius.

1. The diameter of a cylindrical steel rod is 2 cm.
If it is 12 cm long calculate its volume:
(*a*) using $\pi = 3$ (*b*) using $\pi = 3.14$.

2. A cylindrical storage container for flour has a base diameter of 16 cm and a height of 15 cm (all dimensions being internal). Calculate the volume of flour it will hold when full.
(Use $\pi = 3.14$.)

*See the glossary, p.478.

3. A cylindrical storage container for sugar has a base diameter of 13.4 cm and a height of 12.5 cm (all dimensions being internal). Calculate the volume of sugar it will hold when full.
(Use $\pi = 3.14$.)

4. A cylindrical bucket has a diameter of 22 cm and a height of 25 cm (both are internal dimensions).
(*a*) How many cubic centimetres of water can it hold when it is full? (Use $\pi = 3.14$.)
(*b*) Find the capacity of the bucket in litres.

5. A cylindrical milk pan has a diameter of 14 cm and a height of 7 cm (both are internal dimensions).
(*a*) Using $\pi = \frac{22}{7}$ find the number of cubic centimetres of liquid the pan can hold.
(*b*) Find the capacity of the pan in litres.

6. A can has an external diameter of 7.5 cm and a height of 10 cm.
(*a*) Calculate the space taken up by one can. (Use $\pi = 3.14$.)
(*b*) If 12 cans are packed tightly into a box in three rows of four as shown in the sketch, find the minimum internal dimensions of the box.
(*c*) Find the volume of the box.

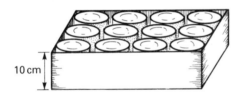

7. A sealed can is made to hold a certain volume, if its height equals its base diameter then less 'tin' is needed to make it.
(*a*) Calculate the volume contained by such a can if its diameter and height both measure 8 cm internally. (Use $\pi = 3.14$.)
(*b*) Measure some tins at home to check whether or not manufacturers have made 'cheaper' tins.
Note that although a manufacturer may save money by making tins where the diameter equals the height and hence uses less 'tin' they may in fact lose money by selling less! Such tins may appear to look smaller than other size tins so people may buy other makes thinking they are getting better value for money.

167

8. In a certain 4-cylinder car, each cylinder has a diameter of 9.3642 cm.

The length of the piston stroke is 6.035 cm.

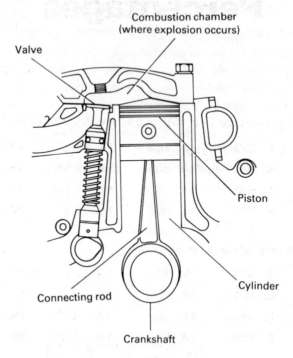

Combustion chamber
(where explosion occurs)

Valve

Piston

Cylinder

Connecting rod

Crankshaft

Calculate the cubic capacity of the car, that is, its engine size in cc's (cubic centimetres).

Note that the cross-sectional area of the cylinder multiplied by the length of the piston stroke gives the number of cc's for one cylinder. Since there are 4 cylinders, the cubic capacity of one cylinder must be multiplied by 4.

Use $\pi = 3.142$ and give your answer to four significant figures.

Note The correct abbreviation for cubic centimetres is cm^3, however when referring to engines cc is normally used.

14 **Percentages**

Exercise 1

A Write each percentage as a vulgar fraction in its simplest form:

1. 50%	**5.** 20%	**9.** 15%	**13.** $66\frac{2}{3}\%$	**17.** $62\frac{1}{2}\%$
2. 25%	**6.** 30%	**10.** 85%	**14.** $2\frac{1}{2}\%$	**18.** $87\frac{1}{2}\%$
3. 75%	**7.** 80%	**11.** 5%	**15.** $7\frac{1}{2}\%$	**19.** $12\frac{1}{2}\%$
4. 10%	**8.** 70%	**12.** $33\frac{1}{3}\%$	**16.** $37\frac{1}{2}\%$	**20.** $22\frac{1}{2}\%$

B Write each percentage as a decimal:

1. 52%	**5.** 44%	**9.** 93%	**13.** $29\frac{1}{2}\%$	**17.** 12.34%
2. 77%	**6.** 32%	**10.** 7%	**14.** $2\frac{1}{2}\%$	**18.** 7.18%
3. 84%	**7.** 59%	**11.** 3%	**15.** 4.9%	**19.** 8.92%
4. 21%	**8.** 41%	**12.** $17\frac{1}{2}\%$	**16.** 5.6%	**20.** 6.74%

Exercise 2

A Find:

1. 50% of:
   (a) £18   (b) £39   (c) £7.50

2. 25% of:
   (a) £30   (b) £27   (c) £3.76

3. 75% of:
   (a) £12   (b) £19   (c) £5.20

4. 30% of:
   (a) £8   (b) £6.40   (c) £87.90

5. 20% of £7

6. 40% of £256

7. 15% of £199

8. 35% of £82

9. 62% of £21.50

10. 38% of £9.50

11. 11% of £3990

12. 6% of £845

13. $12\frac{1}{2}\%$ of:
    (*a*) £60   (*b*) £152   (*c*) £6.80

14. $2\frac{1}{2}\%$ of:
    (*a*) £280   (*b*) £2300   (*c*) £90

15. $33\frac{1}{3}\%$ of:
    (*a*) £5.70   (*b*) £12.75   (*c*) £32.76

16. $22\frac{1}{2}\%$ of:
    (*a*) £100   (*b*) £80   (*c*) £13.20

B Give each answer correct to the nearest penny:

1. 10% of £7.98   **7.** 16% of £287.99

2. 30% of £12.99   **8.** 97% of £26.50

3. 25% of £38.65   **9.** $22\frac{1}{2}\%$ of £5.88

4. 40% of £19.75   **10.** 12.36% of £42.50

5. 50% of £47.23   **11.** 7.82% of £985.75

6. 19% of £81.72   **12.** 8.04% of £4732.96

Exercise 3

1. Out of 4000 people, 35% were male.
    (*a*) What percentage were female?
    (*b*) How many were female?
    (*c*) How many were male?

2. There were 40 people in a room. If 5% of them owned a dog, how many was that?

3. 3% of 500 people did not own a TV set. How many did own a TV set?

4. A sales person earned 4% commission on sales totalling £6500. How much commission was that?

5. The price of a holiday increased by 2%. How much was that increase on a holiday costing £290?

6. A wardrobe costs £121.73 before VAT. How much needs to be added for each of the following VAT rates?
(a) 15%   (b) 12%   (c) $12\frac{1}{2}\%$
(Work to the nearest penny.)

7. Mr Thornton's taxable income came to £5200. If the tax rate was 30%, how much income tax would he pay?

8. A car tyre cost £46.10. If its price increased by 7% (due to inflation!), find the increase in cost.

9. An estate agent sold a house for £39 000. Calculate the commission payable if the rate was:
(a) 1%   (b) $1\frac{1}{2}\%$ (c) $1\frac{1}{4}\%$ (d) $1\frac{3}{4}\%$

10. A car cost £8271 new. If its value depreciated by 36%, calculate:
(a) the amount by which its value depreciated,
(b) its current value.

Expression of One Quantity as a Percentage of Another

£2   is   one-half   of   £4 (since $\frac{2}{4} = \frac{1}{2}$)

so £2   is   50%   of   £4 (since $\frac{1}{2} = 50\%$)

$$\frac{1}{2} = \frac{1}{2} \times 100\% = 50\%$$

£5   is one quarter of   £20 (since $\frac{5}{20} = \frac{1}{4}$)

so £5   is   25% of   £20 (since $\frac{1}{4} \times 100\% = 25\%$)

Exercise 4

e.g.   Write 2 kg as a percentage of 5 kg.

2 kg $= \frac{2}{5}$ of 5 kg

but $\frac{2}{5}$ $= \frac{2}{5} \times 100\% = 40\%$

so 2 kg $= \underline{40\% \text{ of } 5\text{ kg}}$

1. Write £3 as a percentage of £12.

2. Write 5 kg as a percentage of 50 kg.

3. Write 3 ℓ as a percentage of 5 ℓ.

4. Write £7 as a percentage of £35.

5. Write £600 as a percentage of £800.

6. Write £6 as a percentage of £40.

7. Write £700 as a percentage of £2000.

8. Write 3 dollars as a percentage of 60 dollars.

9. Write 9 francs as a percentage of 72 francs.

10. Write £2.50 as a percentage of £100.

11. Write £117 as a percentage of £3000.

12. Write £3.70 as a percentage of £50.

Simple Interest

If you saved £100 for 1 year at a rate of 10% p.a., you would receive 10% of £100 as interest.

The interest for 1 year would be 10% of £100 = $\dfrac{10}{100} \times £100$

so the interest on £100 for 1 year at 10% p.a. = $100 \times \dfrac{10}{100}$ pounds.

The interest on £200 for 1 year at 10% p.a. = $200 \times \dfrac{10}{100}$ pounds

and the interest on £P for 1 year at 10% p.a. = $P \times \dfrac{10}{100}$ pounds.

If the interest rate was r% p.a. instead of 10% p.a.,

the interest on £P for 1 year at r% p.a. = $P \times \dfrac{r}{100}$ pounds

$$= \dfrac{Pr}{100} \text{ pounds.}$$

The interest on £P for 2 years at r% p.a. = $\dfrac{Pr}{100} \times 2$ pounds.

The interest on £P for 5 years at r% p.a. $= \dfrac{Pr}{100} \times 5$ pounds

and the interest of £P for n years at r% p.a. $= \dfrac{Pr}{100} \times n$ pounds

$$= \dfrac{Prn}{100} \text{ pounds.}$$

We obtain the formula:

$$I = \dfrac{Prn}{100}$$

I = the amount of *interest* obtained.

P = the *principal*, which is the amount of money that is saved.

r = the interest *rate*, i.e., the rate per cent per annum.
(*R* is sometimes used to stand for the rate % p.a.)

n = the *number* of years the money is saved.
(*T* is sometimes used instead of *n*, where *T* suggests the *time* in years the money is saved.)

This type of interest, where a fixed sum of money at a fixed rate per cent earns the same amount of interest each year, is called *simple interest*.

(*Note* Normally the amount of interest is added to the original amount saved. In the following year not only is interest earned on the amount of money that was saved but also on the interest that was added to the account. Each year the amount of interest increases since interest is obtained on interest. This type of interest is called *compound interest* and will be dealt with in book 5G.)

Returning to the simple interest formula:

$$I = \dfrac{Prn}{100} \quad \text{or} \quad I = \dfrac{PRT}{100} \quad \text{or}$$

$$\text{Simple Interest} = \dfrac{\text{Principle} \times \text{rate} \times \text{years}}{100}$$

If you save £P and get £I interest, the total amount received, £A, is the sum of the principal and the interest. We obtain the formula:

$$A = P + I \quad \text{or} \quad \text{Amount received} = \text{Principal} + \text{Interest}$$

Exercise 5

A Find the simple interest, I (P = principal, r = rate % p.a. and n = the number of years):

	P	r	n
1.	£100	10%	2
2.	£100	8%	2
3.	£100	12%	3
4.	£200	10%	3
5.	£200	9%	4
6.	£400	5%	3
7.	£500	7%	6
8.	£1000	13%	7
9.	£3000	11%	5
10.	£4600	6%	2

	P	r	n
11.	£250	12%	6
12.	£84	7%	3
13.	£370	15%	8
14.	£980	11%	4
15.	£625	6%	5
16.	£553	4%	3
17.	£2180	9%	2
18.	£5216	8%	7
19.	£1066	$7\frac{1}{2}$%	4
20.	£338	$2\frac{1}{2}$%	5

B Find the simple interest on the following, giving your answers correct to the nearest penny:

1. £417.40 for 2 years at 7% p.a.

2. £625.70 for 3 years at 12% p.a.

3. £118.20 for 3 years at 8% p.a.

4. £35.29 for 4 years at 10% p.a.

5. £79.63 for 5 years at 6% p.a.

6. £2093.70 for 3 years at 9% p.a.

7. £809.50 for 3 years at $7\frac{1}{2}$% p.a.

8. £3197 for 5 years at $8\frac{1}{2}$% p.a.

9. £482.96 for 2 years at $5\frac{3}{4}$% p.a.

10. £83.60 for 4 years at 8.57% p.a.

11. £1654.75 for $2\frac{1}{2}$ years at 4% p.a.

12. £526 for $3\frac{1}{2}$ years at 7.43% p.a.

Exercise 6

A Calculate the total amounts received if:

1. Principal = £200, Interest = £14

2. Principal = £190, Interest = £19

3. Principal = £765, Interest = £61.20

4. Principal = £2980, Interest = £961.05

5. Principal = £39.70, Interest = £2.88

6. Principal = £612.53, Interest = £227.27

B Calculate the total amounts received from the following investments:

1. £100 for 3 years at 7% p.a.

2. £450 for 2 years at 10% p.a.

3. £730 for 5 years at 8% p.a.

4. £2000 for 4 years at 11% p.a.

5. £98 for 5 years at 14% p.a.

6. £847 for 2 years at 9% p.a.

7. £369 for 4 years at 12% p.a.

8. £148 for 6 years at $8\frac{1}{2}$% p.a.

9. £1460 for 3 years at $10\frac{1}{2}$% p.a.

10. £528 for $2\frac{1}{2}$ years at 9% p.a.

C Calculate the total amounts received from the following investments giving each answer correct to the nearest penny:

1. £57.90 for 3 years at 8% p.a.

2. £181.46 for 2 years at 5% p.a.

3. £976.50 for 7 years at 11% p.a.

4. £637.20 for 4 years at 7% p.a.

5. £709.65 for 5 years at 13% p.a.

6. £543 for 5 years at $11\frac{1}{2}$% p.a.

175

7. £106.12 for 3 years at $6\frac{1}{2}$% p.a.

8. £5609.87 for $3\frac{1}{2}$ years at 12% p.a.

9. £895.70 for 6 years at 7.98% p.a.

10. £291.53 for $1\frac{1}{2}$ years at 9.07% p.a.

Percentage Increase and Decrease

Exercise 7

Work out the following using a calculator:

e.g.   115% of £200 = 1.15 × £200 = £230

$$\uparrow$$

since 115% = 1.15

(On a calculator, key in: $\boxed{1}\ \boxed{\cdot}\ \boxed{1}\ \boxed{5}\ \boxed{\times}\ \boxed{2}\ \boxed{0}\ \boxed{0}\ \boxed{=}$)

1. 120% of £800

2. 105% of £94

3. 109% of £342

4. 110% of £12.80

5. 104% of £2487

6. 114% of £591

7. 125% of £270.40

8. 118% of £81.50

Percentage Increase

If £200 is to be increased by 15%,

the increase = 15% of £200 = 0.15 × £200 = £30.

New amount = original amount + increase

$$= £200 \qquad\qquad + £30$$

$$= \underline{£230}$$

Here is a quicker method:
New amount = 115% of £200 = 1.15 × £200 = £230

$$\uparrow$$

(*Note*   New amount = original amount + increase

$$= 100\% \text{ of original amount } +$$

$$15\% \text{ of original amount}$$

$$= 115\% \text{ of original amount})$$

Exercise 8

Increase the following by the percentages given:

e.g. Increase £108 by 2%

New amount = 102% of £108 = 1.02 × £108 = £110.16

1. Increase £100 by 10%.
2. Increase £150 by 30%.
3. Increase £84 by 25%.
4. Increase £600 by 45%.
5. Increase £12 by 18%.
6. Increase £9.50 by 8%.

7. Increase £2800 by 12%.
8. Increase £17.80 by 5%.
9. Increase £45 by 100%.
10. Increase £76 by 225%.
11. Increase £7800 by $12\frac{1}{2}$%.
12. Increase £478 by $7\frac{1}{2}$%.

Percentage Decrease

If £560 is to be decreased by 15%,

New amount = original amount − decrease

= 100% of original amount − 15% of original amount

= 85% of original amount

So all we need to write is:

New amount = 85% of £560 = 0.85 × £560 = £476

Exercise 9

1. Decrease £100 by 10%.
2. Decrease £100 by 15%.
3. Decrease £400 by 7%.
4. Decrease £95 by 20%.
5. Decrease £2600 by 25%.

6. Decrease £8.20 by 5%.
7. Decrease £37 by 2%.
8. Decrease £12.50 by 66%.
9. Decrease £638 by $22\frac{1}{2}$%.
10. Decrease £960 by $2\frac{1}{2}$%.

Exercise 10

Answer the following. Where necessary, give answers to the nearest penny.

1. If the VAT rate is 15%, find the price inclusive of VAT from the pre-VAT prices:
    (*a*) £72   (*b*) 49 p   (*c*) £173.90   (*d*) £25.75   (*e*) £2.48

2. If a bank's interest rate is 7% p.a., calculate how much is in an account after interest has been added if the following amounts were in the account at the beginning of the year:

    (*a*) £59   (*b*) £296   (*c*) £1927   (*d*) £17.80   (*e*) £9.75

3. A finance company charges a flat rate of 12% on loans made for one year. Calculate the total that is repayable on the following loans.

    (*a*) £800   (*b*) £2500   (*c*) £4600   (*d*) £375   (*e*) £195

4. Nigel works in a shop. In a sale, all prices are to be reduced by 12% and Nigel has been asked to work out the new prices and to change the price tickets. Find the sale prices of the following original prices.

    (*a*) £20   (*b*) £12.50   (*c*) £32.90   (*d*) £41.25   (*e*) £57.99

5. The workers at a firm joined a pension scheme and paid 6% of their earnings towards their pension. Calculate their earnings after the pension payment (ignore other deductions), if the total earnings were as follows.

    (*a*) £156   (*b*) £208   (*c*) £320   (*d*) £182.70   (*e*) £219.65

Exercise 11

1. Mr Usher earns £214 per week and pays 9% of these earnings for National Insurance. How much is left after the National Insurance payments have been made?

2. Mrs Ashton earned £810 per month. If she was given a rise of 4%, calculate her new gross monthly earnings.

3. If house values rose by 3%, find the new value of a house that was priced at £37 000.

4. A car worth £8590 depreciated in value by 31% during the year. How much is it now worth?

5. Mr Khan's taxable income was £6370. How much was left after tax if the tax rate was:

    (*a*) 30%?   (*b*) 31%?   (*c*) 29%?   (*d*) 35%?   (*e*) 32%?

Profit and Loss

If a bicycle was bought for £100 then sold for £120 there has been a *profit* of £20.

> Profit = selling price − cost price

However, if the bicycle was bought for £100 then sold for £80, the selling price is less than the cost price and there has been a *loss* of £20.

> Loss = cost price − selling price

These two formulae can also be written as:

> Selling price = cost price + profit
>
> Selling price = cost price − loss

If SP = selling price and CP = cost price, the formulae become:

> Profit = SP − CP SP = CP + profit
> Loss = CP − SP SP = CP − loss

Is £2 profit a good profit? This question cannot be answered without having more information.

Consider the following cases:

1. If something was bought for £1000 then sold for £1002 the profit of £2 is poor. The profit is then 0.2% of the cost price (since £2 is 0.2% of £1000).

2. If something was bought for £100 then sold for £102, once again the profit is £2. It is still a small profit—it is 2% of the cost price (£2 is 2% of £100).

3. If something was bought for £10 then sold for £12, again the profit is £2. The profit is now good—it is 20% of the cost price. (£2 is 20% of £10).

4. If something was bought for £1 then sold for £3, a profit of £2 is exceptionally good and would probably be regarded as unreasonable. It is 200% of the cost price (£2 is 200% of £1).

Writing the profit as a percentage of the cost price makes it easier to recognise how good the profit is.

Note Some firms give the profit (or loss) as a percentage of the selling price. In questions, unless stated otherwise, assume the profit (or loss) is a percentage of the *cost price*.

Note   Profit = % profit × CP

Exercise 12

A Find the selling price if:

	CP	Profit			CP	Loss
1.	£40	20%	**6.**	£50	10%
2.	£2	25%	**7.**	£70	5%
3.	£280	35%	**8.**	£840	2%
4.	£9.35	60%	**9.**	£3.50	8%
5.	£450	23%	**10.**	£34.75	16%

B Answer these. Where necessary give your answer correct to the nearest penny:

1. Find the selling price:
  (a) CP = £69,   profit = 9%
  (b) CP = £720,   profit = 98%
  (c) CP = £290,   profit = 120%
  (d) CP = £81.57,   profit = 200%
  (e) CP = £7.20,   profit = 12.5%
  (f) CP = £900,   profit = $8\frac{3}{4}\%$

2. Find the selling price:
  (a) CP = £140, loss = $2\frac{1}{2}\%$ (b) CP = £200, loss = $6\frac{1}{4}\%$

C Find the selling prices. Where necessary, give answers correct to the nearest penny (ignore VAT).

1. A hairdryer was bought for £11.50 and sold at a profit of 25%.

2. A car was bought for £5000 and sold at a profit of 32%.

180

3. A car was bought for £3800 and sold at a loss of 4%.

4. A jug kettle was bought for £20.97 and sold at a profit of 19%.

5. A power drill was bought for £47.28 and sold at 48% profit.

6. A bicycle was bought for £120 and sold at a loss of 37%.

7. Goods costing £540 were sold at a loss of 14%.

8. Goods costing £268 were sold at a profit of 112%.

Note

$$\% \text{ profit} = \frac{\text{profit}}{\text{CP}} \times 100\%$$

$$\% \text{ loss} = \frac{\text{loss}}{\text{CP}} \times 100\%$$

Exercise 13

A Find the percentage profit (or loss):

	CP	Profit
1.	£30	£3
2.	£40	£10
3.	£100	£20
4.	£70	£21
5.	£20	£1.20

	CP	Loss
6.	£30	£3
7.	£100	£40
8.	£65	£39
9.	£80	£4
10.	£15	£7.50

B Find (*a*) the profit (or loss), (*b*) the percentage profit (or loss):

	CP	SP
1.	£50	£80
2.	£30	£45
3.	£200	£170
4.	£75	£78
5.	£225	£198

	CP	SP
6.	£450	£360
7.	£68	£51
8.	£55	£59.40
9.	£20	£20.60
10.	£15	£16.65

C For each question, calculate:
(a) the profit (or loss), (b) the percentage profit (or loss).

1. Goods costing £85 were sold for £102.

2. A bag costing £12 was sold for £15.

3. A shopkeeper paid £32 for some saucepans then sold them for £36.

4. A car was bought for £8000 then sold for £5600.

5. A kitchen tool set cost £12 then was sold for £12.60.

6. A keyboard costing £180 was sold for £153.

15 Symmetry

Exercise 1   M

1. (a) Copy the triangle and draw its axis of bilateral symmetry.
   (b) What sort of triangle is it?

2. Draw any plane shape that has:
   (a) two axes of bilateral symmetry,
   (b) three axes of bilateral symmetry.

3. Draw any plane shape that has rotational symmetry:
   (a) of order 2,   (b) of order 4,   (c) of order 3.

4. Make three copies of the given diagram.
   Complete each copy so that:
   (a) The first has one axis of bilateral symmetry.
   (b) The second has two axes of bilateral symmetry.
   (c) The third has rotational symmetry of order 3.

5. The given letters are symmetrical. Calculate, for each one, the angle marked with a question mark.

(a)   (b)   (c)   (d)

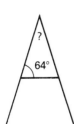

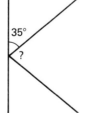
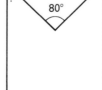
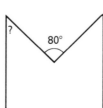
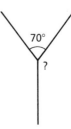

183

6. Draw a pair of axes where the x-values range from 0 to 8 and the y-values range from 0 to 5. Use a scale of 1 cm to 1 unit on both axes. Plot the points O (0, 0), P (5, 0), Q (8, 4), R (3, 4) and join them to form quadrilateral OPQR.
(a) What type of quadrilateral have you drawn?
(b) Draw its axes of bilateral symmetry.

Exercise 2 Properties of Polygons Directly Related to their Symmetries

A The triangle in Exercise 1, question 1 has one axis of bilateral symmetry.

1 Draw a triangle that has no axes of symmetry. What sort of triangle is it?

2. Try to find a triangle with exactly two axes of symmetry. If you find such a triangle, mark its equal angles and equal sides then write its name.

3. Try to find a triangle with exactly three axes of symmetry. If you find such a triangle, mark its equal angles and equal sides then write its name.

B Try to find quadrilaterals with the given number of axes of symmetry. There may be no possible answer, there may only be one possible answer or there may be more than one.
Draw and name each quadrilateral you find.
Mark on each drawing the equal sides and the equal angles.

1. 0   2. 1   3. 2   4. 3   5. 4

C Try to find pentagons with the given number of axes of symmetry. Draw each pentagon you find and mark on each drawing the equal sides and the equal angles.

1. 0   2. 1   3. 2   4 3   5. 4   6. 5

D Try to find hexagons with the given number of axes of symmetry. Draw each hexagon you find and mark on each drawing the equal sides and the equal angles.

1. 0   2. 1   3. 2   4. 3   5. 4   6. 5   7. 6

E Investigate other polygons in the same way as in parts C and D. Write what you notice.

Properties of a Circle Directly Related to its Symmetry

Exercise 3

A 1. Draw any circle and mark its centre with a small cross. (Use a radius of about 30 mm.)

2. In the circle, draw any two chords that are not parallel.

3. Bisect both chords (draw their perpendicular bisectors).

4. Where do the perpendicular bisectors meet?

5. Draw another chord and bisect it to test your answer.

B 1. Draw a circle with a radius between 20 mm and 30 mm. Label its centre O.

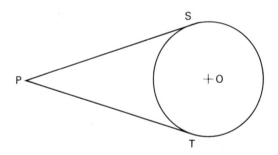

2. Mark a point P somewhere outside the circle.

3. Draw two tangents to the circle from P. Label the points at which the tangents touch the circle as S and T. Measure PS and PT. Write what you notice about these two lengths.

4. Bisect angle SPT (the angle between the tangents). What do you notice about the bisector?

185

C Copy and learn the following:

1. The perpendicular bisector of a chord passes through the centre of a circle.

2. Tangents drawn from one point to a circle are equal in length.

3. The bisector of the angle between two tangents passes through the centre of the circle.

Exercise 4

A Write what you remember about the angle between a tangent and a radius of a circle.

B Calculate the angles labelled with letters (O is the centre of the circle):

1.

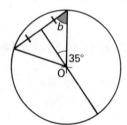

4.

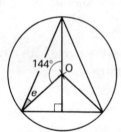

2.

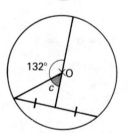

5.

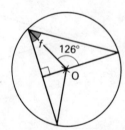

3.

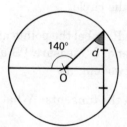

6.

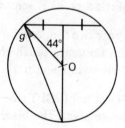

186

7.

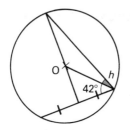

8.

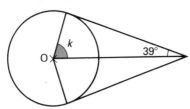

9.

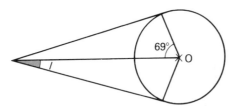

10.

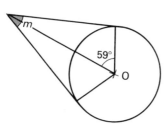

11.

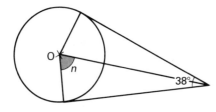

12.

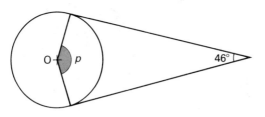

Symmetric Properties of Solid Shapes

Exercise 5

Here are sketches of six different solids:

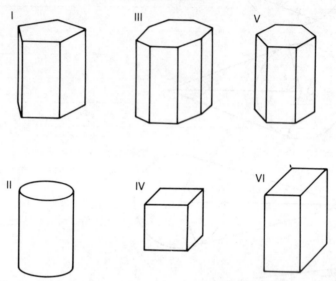

Here are the names of the same six solids but in a different order:

cube, cuboid, pentagonal-based prism, cylinder, hexagonal-based prism, octagonal-based prism

Here is a plan of a child's toy; shapes are posted into the box:

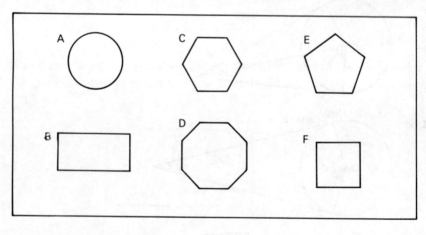

1. Copy the table given below:

Hole	Solid number	Name of solid	Number of ways the solid fits
A			
B			
C			
D			
E			
F			

2. Complete the table by filling in the Roman numeral to show which solid fits exactly through which hole. Also fill in the column to show the name of each solid.

Each solid will fit through its correct hole in a number of different ways. Complete the last column of the table to show the number of different ways each solid fits through each hole.

Planes of Symmetry

A straight line can be cut into two equal parts. The cut gives a *point*.

Point

If a plane shape (that is, a flat shape) has bilateral symmetry, it can be cut into two identical parts. The cut is a *line of symmetry*.

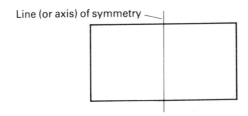

Line (or axis) of symmetry

If a solid can be cut into two identical parts by a straight cut, the solid must be symmetrical.

The cut is a *plane of symmetry*.

The sketches below show three different planes of symmetry for a hexagonal-based prism:

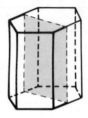 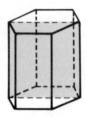 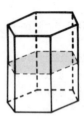

Note Crystals are usually classified according to their symmetry.

Exercise 6

1. (*a*) How many planes of symmetry has a hexagonal-based prism?
   (*b*) If the prism is standing on a hexagonal face, how many of the planes of symmetry are vertical?

2. How many planes of symmetry has:
   (*a*) a cube?　　　　　　　　(*b*) a cuboid?

3. How many planes of symmetry has a triangular-based prism?

4. How many planes of symmetry has:
   (*a*) a square-based pyramid?
   (*b*) a hexagonal-based pyramid?
   (*c*) a triangular-based pyramid?
   (*d*) a regular tetrahedron?

Revision Exercises VIII to XV

Revision Exercise VIII

1. Multiply out:
   (a) $4(x + 5)$
   (b) $2(p - 3)$
   (c) $3(2d + 1)$
   (d) $6(3c - 2)$
   (e) $^-2(3k - 2)$
   (f) $^-4(1 - e)$

2. Factorise:
   (a) $12t + 8$
   (b) $5u - 15$
   (c) $18 - 21m$

3. Find the value of $7 \times 6.8 + 7 \times 3.2$ by factorising.

4. Simplify:
   (a) $4t + 7t - 8t$
   (b) $5k + 2l - 3k$
   (c) $7g - 9g - 2g + 3g + 8g$
   (d) $8m - 7u - 2u - 3m + 4u$

5. Simplify $3gh + 2gk + 4hg - 2gh - gk + 6kg$.

6. Simplify:
   (a) $6 + (u - 2)$
   (b) $3d + (2d + 4)$
   (c) $4k + (10 - k)$
   (d) $10 - (6 - w)$
   (e) $8t - (3t + 9)$
   (f) $5l - (2l - 1)$

7. Simplify:
   (a) $8 + 2(g + 6)$
   (b) $9 + 3(v - 2)$
   (c) $14n + 2(3 - n)$
   (d) $12 - 4(a + 3)$
   (e) $8p - 2(p - 7)$
   (f) $17 - 5(2 + h)$

8. Simplify:
   (a) $3k + 5(2k - 3)$
   (b) $19 - 6(3q - 2)$

9. Multiply out $m(m - 4)$.

10. Multiply out and simplify your answers:
   (a) $(h + 4)(h + 2)$
   (b) $(x - 4)(x + 2)$

191

Revision Exercise IX

1. Sketch a cube.

2. (*a*) How many edges has a square-based pyramid?
  (*b*) How many vertices has a square-based pyramid?
  (*c*) How many faces has a square-based pyramid?

3. Here is the net of a solid:

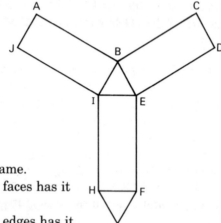

(*a*) Write its name.
(*b*) How many faces has it got?
(*c*) How many edges has it got?
(*d*) How many vertices has it got?
(*e*) In making the solid, edge AB joins edge CB. Which edge does DE join?
(*f*) Which edge does HG join?
(*g*) Which letter meets F at the same vertex?
(*h*) Sketch another possible net of the same solid.

4. Draw a straight line, 5 cm long. Bisect it using a pair of compasses.

5. Draw a straight line, AB, 8 cm in length. Mark a point P above the line and 3 cm from the line. Drop a perpendicular from P to line AB using a pair of compasses.

6. Construct triangle JKL where JK = 8 cm, ∠JKL = 40° and KL = 7 cm. Using a pair of compasses, bisect angle LJK and let the bisector meet KL at M. How long is KM?

1. If you earned £5.20 an hour how much would you earn in:
(*a*) 8 h?   (*b*) 40 h?

2. If £171.50 is paid for 35 h work, how much is that per hour?

3. Mr Norris earned £180.80 gross and deductions totalled £51.95. Find his net wage.

4. Mrs Carr's basic wage was £197.20. If she also earned £43.50 overtime and if deductions totalled £82.80, calculate her net wage.

5. Brenda worked the following hours during a certain week:
Mon 8 h 25 min, Tue 8 h 40 min, Wed 7 h 50 min,
Thur 8 h 20 min, Fri 7 h 20 min, Sat 3 h 25 min.
(*a*) Calculate the total number of hours worked.
(*b*) Calculate her gross earnings at £4.25 an hour.

6. Calculate the overtime earnings:
(*a*) 4 h at double time if the basic rate is £4.60 an hour,
(*b*) 10 h at time and a half if the basic rate is £5.20 an hour.

7. Martyn earns £153 for a basic week of 34 h.
(*a*) Calculate his basic rate per hour.
(*b*) If during a certain week he worked 40 h and overtime was at time and a half, calculate his gross weekly earnings.

8. How much per annum is:
(*a*) £139.70 per week?
(*b*) £745.60 per month?

9. Clive earns £168 per week while Desmond earns £672 per month. Which of them earns the greater amount in a year?

10. Is £100 a week the same amount as £400 a month?
(If not, explain why not.)

11. Nancy earns £8112 p.a. How much is that:
(*a*) per month?   (*b*) per week?

12. Calculate the commission at 4% on goods sold for £875.

13. What commission is earned on sales totalling £3148 if the rate is $2\frac{1}{2}\%$?

14. If Caroline earns 17p for each item she makes, what would she earn for making 620 items?

15. Dennis is paid 3.5p per SM (Standard Minute) and his job is rated at 0.68 SM per unit of work. How much does he earn for completing 1200 units of work.

16. Vivien earned £140 plus a bonus of 10%. Find her total earnings.

Revision Exercise XI

1. Elizabeth saves £3.60 per week. How much is that per year?

2. A bill for £147.08 was paid in two equal instalments. How much was each instalment?

3. Mr Atkinson's quarterly bills for last year were £67.73, £78.88, £44.61 and £57.45. If he had paid regular payments of £23 per month, how much refund should he receive?

4. House insurance costs Mrs Campbell £62.50 for one year. To pay that amount in full, how much does she need to save:
(a) each month?
(b) each week?

5. Which is the better buy, 520 g for 86p or 550 g for 89p?

6. Use the table given on p.137 to find the cost for two people to stay at the Hotel Victoria on full board if their departure date is 5 July.

7. Change £45 into pesetas if the exchange rate is 212 pesetas to 1 pound.

8. Find the cost, to the nearest penny, of an article costing 1.82 francs at an exchange rate of 33.9p to 1 franc.

Revision Exercise XII

1. Which of the following diagrams represents the locus of a valve on a bicycle wheel as the bike travels along a level road?

A.　　　　　　　　B.　　　　　　　　　　　　　C.

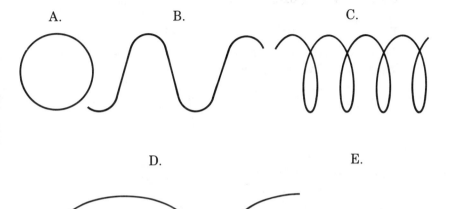

D.　　　　　　　　　　　　　　E.

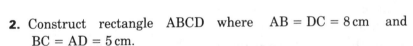

2. Construct rectangle ABCD where AB = DC = 8 cm and BC = AD = 5 cm.
   On BC, mark a point E such that $A\hat{D}E = 70°$. Construct the locus of a point which moves so that it is always equidistant from AB and AD. Let this locus cross DE at F and meet DC at G. Join EG. How long is:
   (a) EF?　　　　　　　　　(b) EG?

3. Draw a pair of axes as shown. Draw the locus of a point that moves so that the sum of its x- and y-co-ordinates is always 7.

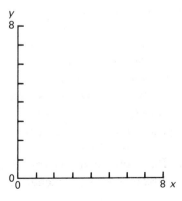

1. A square has a perimeter of 40 cm, find its area.

2. A rectangular garden is 12 m long and $7\frac{1}{2}$ m wide. Find its area.

3. Calculate the area of the parallelogram giving your answer correct to three significant figures.

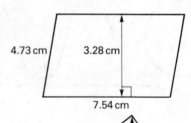

4. The base of a triangle measures 7 cm. If the triangle has an area of 16.1 cm², calculate its perpendicular height.

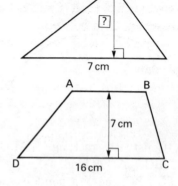

5. In trapezium ABCD, AB is parallel to DC, DC = 16 cm, AB = $\frac{1}{2}$DC and the perpendicular distance between the parallel sides is 7 cm. Calculate the area of the trapezium.

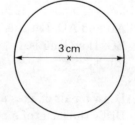

6. (a) Find the circumference of the circle shown. Use Circumference = 3.14 × diameter.
   (b) Find the area of the circle shown. Use Area of circle = 3.14 × (radius)².

7. A bicycle wheel has a diameter of 26 in.
   (a) Calculate the circumference of the wheel.
   (b) Calculate the distance cycled by Beth if the wheel makes 80 rev. (Give the answer in inches correct to the nearest inch.)
   (c) Give the distance cycled in feet correct to the nearest foot (12 in = 1 ft).
   (d) Give the distance cycled in yards correct to the nearest yard (3 ft = 1 yd).

8. The sketch is of a garden measuring 25 m by 15 m.

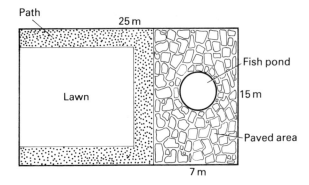

A circular fish pond of radius 1 m is surrounded by a paved area measuring 15 m by 7 m.

A path, 1 m wide, lies on three sides of a lawn as shown.

Calculate:
(a) the area of the fish pond (Use $\pi = 3.14$.),
(b) the area of the paved area that surrounds the fish pond,
(c) the length of the lawn,
(d) the breadth of the lawn,
(e) the area of the lawn,
(f) the area of the path,
(g) the total area of the garden.

9. P and Q lie on a circle whose centre is O. The circle has a radius of 6 cm. $\hat{POQ} = 120°$.
Calculate:
(a) the perimeter of the shaded part.
(b) the area shaded.
(Take π to be 3.)

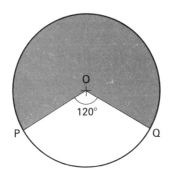

10. A cuboid measures 7 cm by 5 cm by 2 cm.
(a) On squared paper, draw an accurate net of the cuboid.
(b) Find the total surface area of the cuboid.
(c) Calculate the volume of the cuboid.

11. A can of lemonade has a height of 11 cm. The radius of each end is 3.1 cm. Calculate the amount of lemonade it holds. (Use $\pi = 3.142$.)

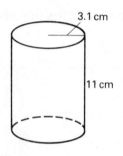

3.1 cm

11 cm

(a) Give your answer in cubic centimetres correct to two significant figures.

(b) Give your answer in millilitres correct to two significant figures.

Revision Exercise XIV

1. (a) Write 60% as a vulgar fraction.
(b) Write 36% as a decimal.
(c) Write $32\frac{1}{2}$% as a decimal.
(d) Write 9% as a decimal.

2. To find $7\frac{1}{2}$% of £4.80 on a calculator, Amanda keyed in:

$\boxed{\cdot}\boxed{0}\boxed{7}\boxed{5}\boxed{\times}\boxed{4}\boxed{\cdot}\boxed{8}$

(a) Which keys should Edward press to find $2\frac{1}{2}$% of £13.60?
(b) What answer should Edward obtain?

3. Barbara paid a bill early and received 5% discount. If the bill was for £350, find the discount.

4. The cash price for a car was £8500. Mrs Eaves decided to buy it.
(a) If she paid a deposit of 30%, calculate the amount of deposit paid.
(b) If the car lost 20% of its value in one year, calculate how much it was worth after a year.

5. Alf got 21 marks out of a possible total of 24. What percentage was that?

6. A two-seater settee costs £300 plus VAT at 15%. Find the total price of the settee including VAT.

7. Find the simple interest on investing:
(a) £500 for 3 years at 6% p.a.,
(b) £350 for 5 years at 8% p.a.

8. Find the simple interest on investing £127.80 for 2 years at 9.81% p.a. Give your answer correct to the nearest penny.

9. £920 was invested for 4 years at 7% p.a. simple interest. Calculate:
  (*a*) the interest received,
  (*b*) the amount received from the investment.

10. (*a*) Increase £300 by 10%.
  (*b*) Decrease £840 by 15%.

11. In a sale, there is a discount of 20%. Find the sale price if the normal price is £18.

12. A shopkeeper buys 40 articles at £3.50 each, then sells them to make a profit of 25%. Calculate the total profit made.

13. Goods costing £60 were sold for £69. Calculate:
  (*a*) the profit, (*b*) the percentage profit.

Revision Exercise XV **M**

1. Copy the following then complete them so that each of the broken lines is an axis of symmetry:

(*a*)   (*c*) 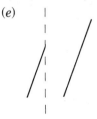   (*e*)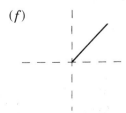

(*b*) 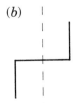   (*d*)   (*f*)

2. (*a*) Draw a pair of axes as shown, then draw the graph of $x = 4$ using a broken line.

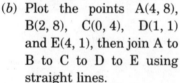

(*b*) Plot the points A(4, 8), B(2, 8), C(0, 4), D(1, 1) and E(4, 1), then join A to B to C to D to E using straight lines.

(*c*) Complete the polygon so that the line $x = 4$ is a line of symmetry.

(*d*) Write the name of the polygon.

(*e*) Calculate the area of the polygon.

3. AB is a chord of a circle whose centre is O. PQ is a straight line passing through O where P lies on the circle and Q is the mid-point of AB. $\hat{POB} = 142°$. Calculate $\hat{QBO}$.

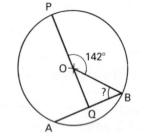

4. PT and PU are tangents to a circle with centre O. If $\hat{TPO} = 28°$ calculate $\hat{UOT}$.

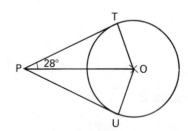

16 **Functions**

Exercise 1

1. Given that $f(x) = 4x$, find:
  (a) $f(3)$ (b) $f(2)$ (c) $f(0)$ (d) $f(^-4)$ (e) $f(^-1)$

2. Given the function $f: x \rightarrow 2x - 3$, carry out these mappings:
  (a) $f: 5 \rightarrow \boxed{?}$ (b) $f: 0 \rightarrow \boxed{?}$ (c) $f: {}^-2 \rightarrow \boxed{?}$ (d) $f: 1 \rightarrow \boxed{?}$

3. If $y = 3x + 7$, find the value of y if:
  (a) $x = 2$ (b) $x = 6$ (c) $x = 0$ (d) $x = {}^-1$ (e) $x = {}^-4$

4. Copy and complete the mapping diagram for the function $g(x) = 5x - 4$:

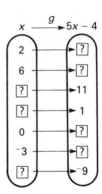

Exercise 2

1. If $h(x) = x^2$, find:
  (a) $h(2)$ (b) $h(0)$ (c) $h(^-3)$ (d) $h(^-5)$ (e) $h(5)$

2. If $y = 3x^2$, find y if:
  (a) $x = 4$ (b) $x = 0$ (c) $x = {}^-5$ (d) $x = {}^-10$ (e) $x = 7$

3. If $f(x) = 2x^2$, find:
  (a) $f(2)$ (b) $f(0)$ (c) $f(4)$ (d) $f(^-3)$ (e) $f(^-6)$

4. If $y = 4x^2$, find y if:
  (a) $x = 2$ (b) $x = 1$ (c) $x = 0$ (d) $x = {}^-3$ (e) $x = {}^-5$

201

5. If $k(x) = 6x^2$ find:

(a) $k(0)$ (b) $k(3)$ (c) $k(5)$ (d) $k(^-1)$ (e) $k(^-3)$

6. Copy and complete the mapping diagram for the function $f(x) = 5x^2$:

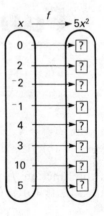

7. Copy and complete the table for the function $y = 8x^2$:

x	$^-6$	$^-5$	$^-4$	$^-3$	$^-2$	$^-1$	0	1	2	3	4	5	6
$y = 8x^2$	?	200	?	?	?	?	?	?	?	72	?	?	?

Exercise 3

Copy and complete the mapping diagram:

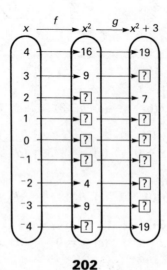

202

Note that for this mapping the middle step may be missed out.

The mapping becomes: which is the same as:

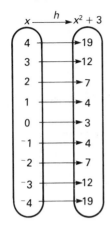   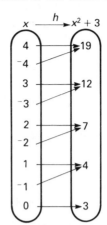

Compare the two mapping diagrams above to see how the second one was obtained. Note that in the second mapping diagram, numbers are written only once in the set.

The domain is $\{^-4, ^-3, ^-2, ^-1, 0, 1, 2, 3, 4\}$.

The second set $\{3, 4, 7, 12, 19\}$ is called the range.*

This diagram also shows the mapping $h: x \rightarrow (x^2 + 3)$:

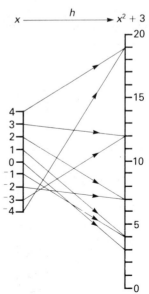

Exercise 4

Draw mapping diagrams for the following functions. Use the domain given:

1. $f(x) = x^2 + 5$ domain $= \{^-3, 3, ^-2, 2, ^-1, 1, 0\}$

2. $g(x) = x^2 - 1$ domain $= \{^-5, 5, ^-2, 2, 0\}$

3. $h(x) = x^2 - 4$ domain $= \{^-5, ^-4, ^-3, ^-2, ^-1, 0, 1, 2, 3\}$

4. $k(x) = x^2 + 8$ domain $= \{^-3, ^-2, ^-1, 0, 1, 2, 3\}$

The function $f(x) = x^2 + 3$ can be written as an equation:
$$y = x^2 + 3.$$

A table can be produced:

x	$^-4$	$^-3$	$^-2$	$^-1$	0	1	2	3	4
x^2	16	9	4	1	0	1	4	9	16
$y = x^2 + 3$	19	12	7	4	3	4	7	12	19

(Compare the table above with the mapping diagram on p.203.)

Ordered pairs (pairs of co-ordinates) can also be produced:

e.g. $(^-4, 19), (^-3, 12), (^-2, 7), (^-1, 4), (0, 3), (1, 4), (2, 7), (3, 12),$
$(4, 19)$

Exercise 5   M

A Copy and complete the following tables for the functions given:

1. $y = x^2 + 1$

x	$^-4$	$^-3$	$^-2$	$^-1$	0	1	2	3	4
x^2								9	
$y = x^2 + 1$								10	

2. $y = x^2 - 2$

x	$^-3$	$^-2$	$^-1$	0	1	2	3	4	5
x^2		4							
$y = x^2 + 2$		2							

3. $y = x^2 + 7$

x	$^-5$	$^-4$	$^-3$	$^-2$	$^-1$	0	1	2	3
x^2		16							
$y = x^2 + 7$		23							

4. $y = x^2 - 8$

x	$^-4$	$^-3$	$^-2$	$^-1$	0	1	2	3	4
x^2		9							
$y = x^2 - 8$		1							

5. $y = x^2 + 4$

x	$^-5$	$^-4$	$^-3$	$^-2$	$^-1$	0	1	2	3
x^2					1				
$y = x^2 + 4$					5				

B Copy and complete the following ordered pairs for the functions given:

1. $y = x^2 + 2$
(0, ?), (1, ?), (4, ?), ($^-$2, ?), ($^-$3, ?), (3, ?), (6, ?)

2. $y = x^2 - 3$
(0, ?), (3, ?), ($^-$3, ?), ($^-$2, ?), (2, ?), (1, ?), (5, ?)

3. $y = x^2 + 10$
(0, ?), (2, ?), (5, ?), ($^-$5, ?), (1, ?), ($^-$1, ?), ($^-$6, ?)

4. $y = x^2 - 6$
(0, ?), (4, ?), ($^-$1, ?), ($^-$3, ?), (6, ?), ($^-$5, ?), (7, ?)

205

1. A function is defined as $f(x) = 3x + 3$.
   (a) Find $f(4)$.
   (b) Evaluate $f(^-2)$.
   (c) Find x when $f(x) = 9$.
   (d) Find x when $f(x) = 18$.

2. A function is defined by the equation $y = 2x - 9$.
   (a) Find y when $x = 6$.
   (b) Find y when $x = 2$.
   (c) Find x when $y = 5$.
   (d) Find x when $y = 1$.

3. A function is defined as $g(x) = 4x - 2$.
   (a) Evaluate $g(1)$.
   (b) Evaluate $g(5)$.
   (c) Find the value of $g(^-2)$.
   (d) Find x when $g(x) = 6$.
   (e) Find x when $g(x) = 14$.
   (f) Find x when $g(x) = {}^-2$.

4. If $y = 4x + 5$, find:
   (a) y when $x = 0$
   (b) y when $x = 4$
   (c) y when $x = {}^-2$
   (d) x when $y = 9$
   (e) x when $y = 17$
   (f) x when $y = {}^-7$

5. A function is defined as $f(x) = x^2 + 4$.
   (a) Evaluate $f(2)$.
   (b) Find the value of $f(5)$.
   (c) Evaluate $f(4)$.
   (d) Find the value of $f(^-4)$.
   (e) Find two values of x if $f(x) = 13$.

6. If $y = x^2 - 5$, find:
   (a) y if $x = 1$
   (b) y if $x = 5$
   (c) y if $x = {}^-4$
   (d) y if $x = {}^-2$
   (e) two values of x if $y = 4$
   (f) x if $y = {}^-5$

7. If $h(x) = x^2 - 7$, find:
   (a) $h(3)$
   (b) $h(0)$
   (c) $h(2)$
   (d) $h(^-2)$
   (e) two values of x if $h(x) = 9$
   (f) two values of x if $h(x) = {}^-6$

8. If $y = x^2 + 6$, find:
   (a) y if $x = 0$
   (b) y if $x = 1$
   (c) y if $x = 4$
   (d) y if $x = {}^-5$
   (e) two values of x if $y = 10$
   (f) two values of x if $y = 42$

1. (*a*) Copy and complete the mapping diagram for the function
$y = (3x + 5)/2$:

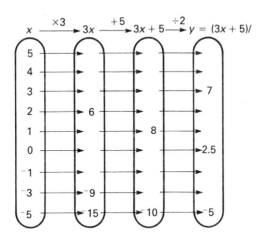

(*b*) The mapping diagram above can be drawn in one stage as shown here. Copy and complete this mapping diagram.

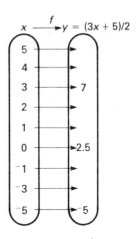

(*c*) Copy and complete the table for the function $y = (3x + 5)/2$.

x	$^-5$	$^-3$	$^-1$	0	1	2	3	4	5
$3x$	$^-15$					6			
$3x + 5$	$^-10$				8				
$y = (3x + 5)/2$	$^-5$			2.5			7		

2. (a) Copy and complete the mapping diagram for the function $y = 2(x + 3)/5$.

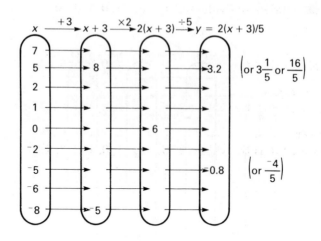

$\left(\text{or } 3\frac{1}{5} \text{ or } \frac{16}{5}\right)$

$\left(\text{or } \frac{-4}{5}\right)$

(b) Copy and complete this one-stage mapping diagram for the function $y = 2(x + 3)/5$:

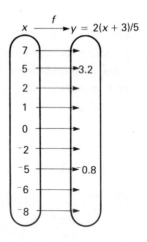

(c) Copy and complete the table for the function $y = 2(x + 3)/5$:

x	$^-8$	$^-6$	$^-5$	$^-2$	0	1	2	5	7
$x + 3$	$^-5$							8	
$2(x + 3)$					6				
$y = 2(x + 3)/5$			$^-0.8$					3.2	

208

3. Draw a mapping diagram and a table for each of the following functions:

(a) $y = 3(x + 1)/4$

where x takes the values $\{^-5, \,^-3, \,^-1, 0, 1, 3, 5\}$.

(b) $y = \dfrac{3(x - 2)}{2}$

where x takes the values $\{^-6, \,^-4, \,^-2, 0, 2, 4, 6, 8\}$.

(c) $y = \frac{2}{3}(x - 3)$

where x takes the values $\{^-6, \,^-3, \,^-2, 0, 1, 3, 6, 9\}$.

(d) $y = (2x - 3)/2$

where x takes the values $\{^-3, \,^-2, \,^-1, 0, 1, 2, 3, 4\}$.

17 **Formulae**

Exercise 1

A **1.** Use $P = 4l$ to find P when $l = 3.4$.

2. Use $l = \dfrac{P}{4}$ to find l when $P = 264$.

3. Use $V = lbh$ to find V when $l = 9$, $b = 4$ and $h = 3$.

4. Use $P = 2(l + b)$ to find P when $l = 6.9$ and $b = 2.5$.

5. Use $P = 2l + 2b$ to find P when $l = 8$ and $b = 4$.

6. Use $l = \frac{1}{2}(P - 2b)$ to find l when $P = 32$ and $b = 7$.

7. Use $v = u + at$ to find v if $u = 0$, $a = 9.8$ and $t = 3$.

8. Use $A = \frac{1}{2}Dd$ to find A if $D = 8.6$ and $d = 4$.

B **1.** The area of the four walls of a room is given by the formula $A = Ph$ where P is the perimeter of the room and h is its height. Find the area of the walls of a room of height 2.4 m if the perimeter measures 18 m.

2. The formula $C = \frac{5}{9}(F - 32)$ can be used to change degrees Fahrenheit (°F) into degrees Celsius (°C). Use it to change 86 °F into degrees Celsius.

3. A formula used in science is:

$$\text{Power} \;=\; \text{voltage} \;\times\; \text{current}$$
$$\text{(watts, W)} \quad \text{(volts, V)} \quad \text{(amperes, A)}$$

Use the formula to find the power of a 12 V car headlamp bulb if it takes a current of 1.5 A.

4. The change in potential energy can be found by using the formula:

$$PE = mgh \qquad \text{where } g = \text{acceleration due to gravity (m/s}^2\text{)},$$
$$m = \text{mass (kg)} \quad \text{and } h = \text{height (m)}$$

Use the formula to find the increase in potential energy of a body of mass 60 kg that is moved vertically upwards through a distance of 8 m. $g = 9.8 \, \text{m/s}^2$ and the potential energy is in joules.

5. $V = Ah$ gives the volume of a prism of perpendicular height h units if the area of its base is A square units.
Find the volume of a prism which has a base area of $4.6 \, \text{cm}^2$ and a perpendicular height of 3 cm.

6. $V = \frac{1}{3}Ah$ gives the volume of a pyramid with base area A square units and perpendicular height h units. Find the volume of a pyramid with base area $6 \, \text{cm}^2$ and perpendicular height 4.5 cm.

C 1. Use $A = l^2$ to find A when:
   (a) $l = 7$ \qquad\qquad (b) $l = 4.5$

2. Use $W = 8d^2$ to find W when:
   (a) $d = 2$ \qquad\qquad (b) d $= 0.6$

3. Use $s = 5t^2$ to find s when:
   (a) $t = 3$ \qquad\qquad (b) $t = 1.5$

4. Use $s = 4.9t^2$ to find s when:
   (a) $t = 2$ \qquad\qquad (b) $t = 3.5$

5. Use $V = 4l^3$ to find V when $l = 3$.

6. If $t = \sqrt{\frac{s}{5}}$ find t when $s = 80$.

7. Use $r = \sqrt{\frac{A}{3}}$ to find r when $A = 75$.

8. Use $l = \sqrt{\frac{V}{h}}$ to find l when $V = 54$ and $h = 6$.

9. Use $l = \sqrt{\frac{3V}{h}}$ to find l when $V = 3$ and $h = 4$.

10. Use $r = \sqrt{\frac{A}{4\pi}}$ to find r when $A = 452$ and $\pi = 3.14$.

Exercise 2

1. How many miles does a train travel:
   (a) in 3 h at a steady 60 m.p.h.?
   (b) in h h at a steady 60 m.p.h.?
   (c) in h h at a steady v m.p.h.?

2. (a) I travelled 8 km on the bus and 15 km on the train. How far is that altogether?
   (b) I travelled b km on the bus and t km on the train. How far is that altogether?

3. (a) Tickets cost £3 each. How many can I buy for £15?
   (b) Tickets cost £x each. How many can I buy for £15?
   (c) Tickets cost £x each. How many can I buy for £y?

4. (a) 6 books cost £5 each. Find the total cost.
   (b) 6 books cost £m each. Find the total cost.
   (c) n books cost £m each. Find the total cost.

5. (a) Mrs Young works 40 h in a week. After working 26 h, how many more hours must she work?
   (b) Mr Varley works 40 h in a week. After working k h, how many more hours must he work?
   (c) Mr Sampson works t h in a week. After working k h, how many more hours must he work?

6. (a) How many hours are there in 180 min?
   (b) How many hours are there in m min?

7. How many days are there in g weeks?

8. How many weeks are there in h days?

Exercise 3

Each of the following formulae is given in words.
Write an expression for each one:

1. The volume of a cuboid, V cm³, can be found by multiplying together its length l cm, its breadth b cm and its height h cm.
   Use the formula to find the volume of a cuboid of length 9 cm, breadth 4 cm and height 3 cm.

2. The number of sides, n, of a regular polygon, can be found by dividing 360° by the size of one of its exterior angles, E, measured in degrees.
Use the formula to find the number of sides of a regular polygon with exterior angles of 20°.

3. The number of seconds, x, in y min, can be found by multiplying the number of minutes by 60.
Write a formula, then use it to find the number of seconds in 8 min.

4. Write a formula for u in terms of w, if w is the number of minutes in u h.
Use the formula to find the number of hours in 240 min.

5. The length, l cm, of a rectangle with area A cm^2, can be found by dividing the area by the breadth b cm.
Use the formula to find the length of a rectangle of area 37 cm^2 and breadth 5 cm.

6. The angles of a triangle are $a°$, $b°$ and $c°$. By subtracting the sum of two of the angles from 180°, the third angle can be found. Write three formulae, one to find a, one to find b and the other to find c.
Use one of the formulae to find the third angle of a triangle where two of the angles measure 49° and 87°.

7. The weight, W newtons, of a body can be written as the product of its mass, m kg, and the acceleration due to gravity, g m/s^2.
Find the weight of a body of mass 67 kg if the acceleration due to gravity is 9.8 m/s^2.

8. By Ohm's law the resistance, R ohms, of a conductor, can be found by dividing its potential difference, V volts, by the current, I amperes.
Find the resistance of an electric bulb connected to a 240 V supply if it takes a current of 0.25 A.

Exercise 4

1. A bath is being filled at a rate of 7 ℓ/min.
   C ℓ of water pours into the bath in t min.
   (a) Make a table showing t and C where t varies from 0 to 7 min.

t (min)	0	1	2	3	4	5	6	7
C (ℓ)	0	7						49

   (b) Plot a graph of C against t. For t use a scale of 1 cm to 1 min and for C use a scale of 1 cm to 5 ℓ.
   (c) Find a formula for C in terms of t.

2. The diagrams show some 1 cm cubes:

    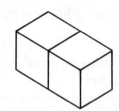 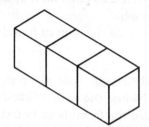

   If n stands for the number of cubes and A cm² is the total surface area of a shape,

   when $n = 1$, $A = 6$
   when $n = 2$, $A = 10$ Check these.
   when $n = 3$, $A = 14$

   and so on.

   (a) Make out a table for n and A (take n from 1 to 8).

Number of cubes, n	1	2	3	4	5	6	7	8
Surface area, A (cm²)	6	10	14	?	?	?	?	

   (b) Plot a graph of A against n.
   Let n be on the horizontal axis and use a scale of
   1 cm to 1 cube. A should be on the vertical axis. Use a scale
   of 1 cm to 1 cm² for A.
   (c) Try to find a formula for A in terms of n. The formula looks like $A = \boxed{?}\, n + \boxed{?}$. (Find the two missing numbers.)
   (d) What is the value of A when $n = 40$?

Exercise 5

1. The right-angled, scalene triangle shown, has sides of x cm, y cm and z cm.
   Write a formula for:

   (a) its perimeter, P cm,
   (b) its area, A cm².

2. Line AB is l cm long:
   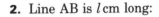

   It is divided into two parts. One part is a cm long and the other part is b cm long.
   Write a formula giving:
   (a) l in terms of a and b,
   (b) a in terms of l and b,
   (c) b in terms of l and a.

3. Give length d in terms of c, e and l:

   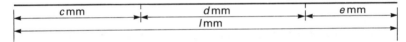

4. (a) How long is WX?
   (b) Write a formula for the perimeter, P cm, of rectangle WXYZ.
   (c) Write a formula for the area, A cm², of rectangle WXYZ.
   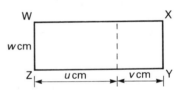

5. Rectangle PQRS has been divided into two smaller rectangles:
   (a) How long is SM?
   (b) Write a formula giving area B m² in terms of p, q and l.
   (c) Write a formula giving area A m² in terms of p, q and l.
   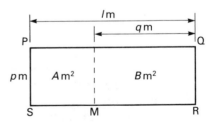

6. A plan of a room is shown. Write a formula for its perimeter, P m.

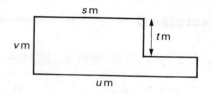

7. Write formulae in terms of k, l, m and n (using some of these letters or all of them) for:

(a) area A,

(b) area B,

(c) the total area of the shape.

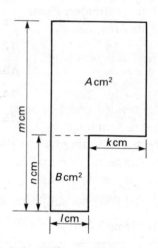

8. A piece of card measuring x cm by 7 cm has a 2 cm square cut from each corner. It is then folded to make a box.

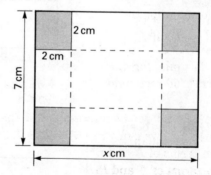

(a) What is the breadth of the box?

(b) What is the length of the box?

(c) Write a formula to find the volume, V cm^3, of the box, giving V in terms of x.

(d) Find the volume if $x = 5$.

(e) Find x if the volume of the box = 24 cm^3.

Exercise 6

A Find x in terms of the other letters:

1. $x + 5 = g$

2. $x + 8 = h$

3. $x + u = 6$

4. $x + t = 14$

5. $x + v = w$

6. $2 + x = k$

7. $m + x = 9$

8. $b + x = c$

9. $x - 9 = e$

10. $x - q = 8$

11. $x - y = 15$

12. $x - a = b$

13. $6x = d$

14. $4x = f$

15. $lx = 24$

16. $nx = p$

17. $\dfrac{x}{3} = d$

18. $\dfrac{x}{5} = r$

19. $\dfrac{x}{a} = 4$

20. $\dfrac{x}{z} = c$

B

1. HP price = loan + interest. Rewrite this, making:
   (a) the loan the subject of the formula,
   (b) the interest the subject of the formula.

2. $a + b = 360$. Rewrite this such that:
   (a) a is the subject of the formula,
   (b) b is the subject of the formula.

3. $A\hat{B}C = 48°$, so $x + y = 48$.
   (a) Find x in terms of y.
   (b) Find y in terms of x.

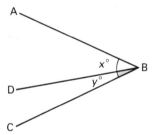

4. If you invest £P and get £I interest and if the amount in your account is then £A, we can write $A = P + I$.
   (a) Find P in terms of A and I.
   (b) Find I in terms of A and P.

5. The radius of curvature of a concave mirror is twice the focal length. This can be written as $r = 2f$. Find f in terms of r.

6. The area of a parallelogram is given by $A = bh$.
   (a) Find b in terms of A and h.
   (b) Find h in terms of A and b.

7. A formula used in science is $F = ma$ where F stands for force, m for the mass of a body and a for the acceleration.
(a) Find m in terms of F and a.
(b) Find a in terms of F and m.

8. The formula $M = 60H$ can be used to change hours into minutes. Rewrite it to change minutes into hours.

Exercise 7

Rewrite each of these formulae as instructed:

1. $A = 4lb$
   (a) $l = \boxed{?}$
   (b) $b = \boxed{?}$

2. $V = lbh$
   (a) $l = \boxed{?}$
   (b) $b = \boxed{?}$
   (c) $h = \boxed{?}$

3. $A = \pi rl$
   (a) $r = \boxed{?}$
   (b) $l = \boxed{?}$

4. $P = mgh$
   (a) $m = \boxed{?}$
   (b) $g = \boxed{?}$
   (c) $h = \boxed{?}$

5. $pV = RT$
   (a) $p = \boxed{?}$
   (b) $R = \boxed{?}$
   (c) $V = \boxed{?}$
   (d) $T = \boxed{?}$

6. $Fy = mgx$
   (a) $F = \boxed{?}$
   (b) $y = \boxed{?}$
   (c) $m = \boxed{?}$
   (d) $g = \boxed{?}$
   (e) $x = \boxed{?}$

7. $f = \dfrac{V}{4l}$ $V = \boxed{?}$

8. $m = \frac{5}{8}k$ $k = \boxed{?}$

9. $A = \dfrac{bh}{2}$
   (a) $b = \boxed{?}$
   (b) $h = \boxed{?}$

10. $r = \dfrac{\pi d}{180}$
   (a) $d = \boxed{?}$
   (b) $\pi = \boxed{?}$

11. $V = \frac{1}{3}Ah$
   (a) $A = \boxed{?}$
   (b) $h = \boxed{?}$

12. $I = \dfrac{Prn}{100}$

    (a) $n = \boxed{?}$

    (b) $P = \boxed{?}$

13. $I = \dfrac{PRT}{100}$

    (a) $R = \boxed{?}$

    (b) $T = \boxed{?}$

14. $F = \dfrac{IHl}{10}$

    (a) $I = \boxed{?}$

    (b) $H = \boxed{?}$

    (c) $l = \boxed{?}$

Exercise 8

Rewrite as instructed:

1. $a + b + c = 180$
    (a) $a = \boxed{?}$ (b) $b = \boxed{?}$

2. $R = r_1 + r_2 + r_3$
    (a) $r_1 = \boxed{?}$ (b) $r_3 = \boxed{?}$

3. $x + y + z = 90$
    (a) $y = \boxed{?}$ (b) $z = \boxed{?}$

4. $l + m + n = 24$
    (a) $l = \boxed{?}$ (b) $n = \boxed{?}$

5. $m + k + t = u$
    (a) $k = \boxed{?}$ (b) $t = \boxed{?}$

6. $r + l + m = g$
    (a) $r = \boxed{?}$ (b) $m = \boxed{?}$

Exercise 9

A Find x:

1. $^-x = ^-5$

2. $^-x = ^-4$

3. $^-x = ^-7$

4. $^-x = ^-a$

5. $^-x = ^-d$

6. $^-x = 3$

7. $^-x = 6$

8. $^-x = m$

9. $^-x = y$

10. $^-x = k$

11. $8 - x = 3$

12. $4 - x = 1$

13. $12 - x = 5$

14. (a) $7 - x = 4$

    (b) $d - x = 4$

    (c) $d - x = f$

15. (a) $9 - x = 5$

    (b) $9 - x = g$

    (c) $h - x = g$

16. $5 = 13 - x$

17. $35 = 90 - x$

18. (a) $2 = 6 - x$

    (b) $2 = t - x$

    (c) $v = t - x$

B 1. $n - x = 12$
   (a) Find n.   (b) Find x.
2. $b - a = c$
   (a) Find b.   (b) Find a.
3. $z - u = e$
   (a) Find z.   (b) Find u.
4. $s - k = 5$
   (a) Find s.   (b) Find k.
5. $p - r = w$
   (a) Find p.   (b) Find r.

C 1. $a - x = 3$, find x.   5. $q = 90 - x$, find x.
2. $8 - x = l$, find x.   6. $y = 180 - z$, find z.
3. $y - t = m$, find t.   7. $c - t = e$, find t.
4. $h - v = 20$, find v.   8. $v = g - h$, find h.

Exercise 10

Rewrite as instructed:

1. $\dfrac{h}{d} = 24$
   (a) $h = \boxed{?}$ (b) $d = \boxed{?}$

2. $\dfrac{s}{m} = 60$
   (a) $s = \boxed{?}$ (b) $m = \boxed{?}$

3. $\dfrac{d}{r} = 2$
   (a) $d = \boxed{?}$ (b) $r = \boxed{?}$

4. $\dfrac{A}{b} = h$
   (a) $A = \boxed{?}$ (b) $b = \boxed{?}$

5. $I = \dfrac{V}{R}$
   (a) $V = \boxed{?}$ (b) $R = \boxed{?}$

6. $v = \dfrac{s}{t}$
   (a) $s = \boxed{?}$ (b) $t = \boxed{?}$

7. $P = \dfrac{F}{A}$
   (a) $F = \boxed{?}$ (b) $A = \boxed{?}$

8. $D = \dfrac{M}{V}$
   $V = \boxed{?}$

9. $\dfrac{p}{l} = 4$
   $l = \boxed{?}$

10. $T = \dfrac{1}{f}$
   $f = \boxed{?}$

Exercise 11

A Solve for x:

1. $2x + 8 = 14$	**5.** $5x + 7 = 22$	**9.** $6x - 7 = 17$
2. $3x + 5 = 17$	**6.** $5x - 7 = 13$	**10.** $10x + 4 = 54$
3. $2x - 2 = 10$	**7.** $4x - 6 = 2$	**11.** $3x + 6 = 24$
4. $3x - 7 = 5$	**8.** $7x + 1 = 15$	**12.** $2x - 9 = 5$

B Find x:

1. (a) $2x - 7 = 3$
   (b) $2x - 7 = a$
   (c) $2x - t = 3$
   (d) $2x - t = a$

2. (a) $3x + 1 = 19$
   (b) $3x + 1 = u$
   (c) $3x + v = 19$
   (d) $3x + v = u$

3. (a) $3x - 2 = 4$
   (b) $3x - 2 = c$
   (c) $nx - 2 = 4$
   (d) $nx - 2 = c$

4. (a) $4x + 8 = 12$
   (b) $kx + 8 = 12$
   (c) $kx + 8 = p$
   (d) $kx + d = 12$

5. $2x + 9 = h$
6. $5x - 2 = b$
7. $3x + t = 9$
8. $4x - l = 10$
9. $bx + 4 = 7$
10. $gx - 9 = 11$
11. $ux + 6 = w$
12. $zx + p = q$

C 1. $v = u + at$, find:
   (a) u (b) t (c) a

2. $s = 180n - 360$, find n.

3. $l = m + pn$, find:
   (a) m (b) p (c) n

4. $P = 2l + b$, find:
   (a) b (b) l

5. $P = 2l + 2b$, find:
   (a) b (b) l

18 Transformations

Exercise 1

1. Describe completely the single transformation that would move:
  (a) triangle A to B,
  (b) triangle A to C,
  (c) triangle B to C,
  (d) triangle A to D,
  (e) triangle A to E.

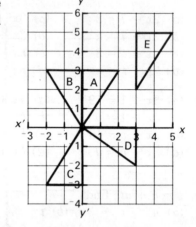

2. (a) Draw a pair of axes as shown. x-values should range from ⁻10 to ⁺10 and y-values from ⁻8 to ⁺8. Use a scale of 1 cm to 1 unit on both axes.

  (b) Copy trapezium A.

  (c) Reflect shape A in the x-axis and label the image, B.

  (d) Rotate shape A through 180° about the origin and label the image, C.

  (e) Describe completely the transformation that would map trapezium B on to trapezium C.

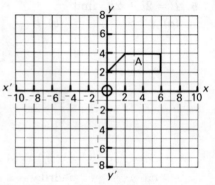

1. (a) Draw a pair of axes as shown. The x-values should range
   from ⁻8 to ⁺8 and the y-values from ⁻10 to ⁺10. (The squares
   shown represent 1 cm squares.)

   (b) Copy the given L-shape and label it L.

   (c) Draw the image of shape L after a reflection in the y-axis.
   Label the image, Y.

   (d) Draw the image of shape L after a rotation through 90°
   anticlockwise about the origin. Label the image, R.

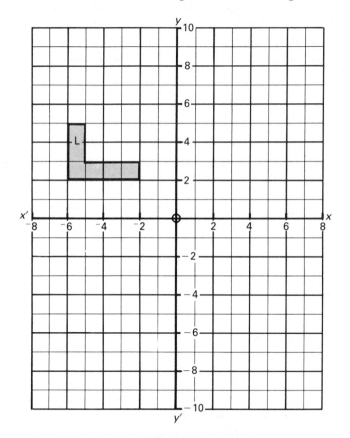

2. Using the same pair of axes as for question 1:

   (a) Plot and label points A(3, ⁻6), B(6, ⁻6), C(5, ⁻8) and
   D(2, ⁻8).

   (b) Join A to B to C to D to A using straight lines. Write the
   name of the quadrilateral obtained.

(c) If A'($^-$7, $^-$7) is the image of A under a translation, plot A' then mark the positions of B', C' and D', the images of B, C and D, under the same translation. Write their pairs of co-ordinates. Join A' to B' to C' to D' to A'.

(d) Reflect shape ABCD in the x-axis. Label the image position, X.

(e) Draw the image of shape ABCD after a rotation through 180° about the origin. Label the image position, N.

3. Draw another pair of axes. The x-values should range from $^-$8 to $^+$8 and the y-values should range from $^-$5 to $^+$5.

(a) Plot the points P(4, 0), Q(4, 2), R($^-$2, $^-$4), S($^-$6, 0) and join them in that order using straight lines.

(b) Now reflect the shape in the x-axis and draw its image.

Exercise 3

Answer the following without plotting any points:

1. The given points are reflected in the x-axis. Find their images.
   (a) (4, 3)   (b) ($^-$2, 5)   (c) ($^-$3, $^-$6)   (d) (7, 0)

2. The given points are reflected in the y-axis. Find their images.
   (a) (2, 2)   (b) ($^-$3, 5)   (c) (1, $^-$3)   (d) (0, 6)

3. The given points are rotated through 180° about the origin. Find their images.
   (a) (2, 4)   (b) ($^-$3, 4)   (c) ($^-$2, $^-$5)   (d) (6, $^-$2)

4. The given points are rotated 90° clockwise about the origin. Find their images.
   (a) (5, 4)   (b) (3, $^-$2)   (c) ($^-$4, 1)   (d) ($^-$3, $^-$5)

5. (a) The point ($^-$3, 2) is translated 4 units to the right (parallel to the x-axis) and 5 units upwards (parallel to the y-axis). Find its image.

   (b) The point (5, $^-$3) is translated 1 unit to the left (parallel to the x-axis) and 2 units downwards (parallel to the y-axis). Find its image.

   (c) Find the image of ($^-$4, 0) if it is translated 3 units to the left (parallel to the x-axis) and 4 units downwards (parallel to the y-axis).

Copy the following shapes then reflect them in the lines labelled m:

1.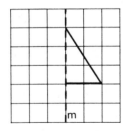

4.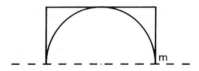

2.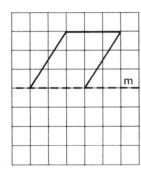

5.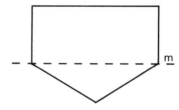

3.

6.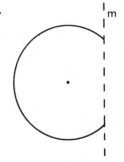

Reflections in Lines Parallel to the Axes

Exercise 5

In the diagram given, the lines $y = 3$ and $x = {}^-2$ have been drawn.

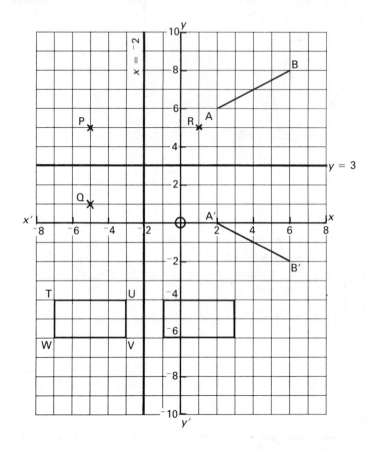

1. The point P has been reflected in the line $y = 3$ giving the image Q.
   (a) What are the co-ordinates of P?
   (b) What are the co-ordinates of Q?

2. What are the co-ordinates of R if R is the image of P after a reflection in the line $x = {}^-2$?

3. The line AB has been reflected in $y = 3$.
Write the co-ordinates of:
(a) A   (b) A′   (c) B   (d) B′

4. Rectangle TUVW is reflected in $x = {}^-2$. T is $({}^-7, {}^-4)$, U is $({}^-3, {}^-4)$, V is $({}^-3, {}^-6)$, and W is $({}^-7, {}^-6)$.
Find the co-ordinates of the vertices of the image of TUVW.

Exercise 6

Draw a pair of axes as for Exercise 5.

1. Draw the line $x = 4$.

2. Draw the line $y = {}^-1$.

3. Plot the point A(6, 8) then reflect it in the line $x = 4$.
Label the image A′. What are the co-ordinates of A′?

4. Plot the points B(1, 6) and C(4, 7) then join them to give a straight line. Reflect BC in the line $x = 4$ and label the image B′C′.
Write the co-ordinates of:
(a) B′   (b) C′

5. Reflect BC in the y-axis and label the image B″C″.
Write the co-ordinates of:
(a) B″   (b) C″

6. If $D = ({}^-5, 4)$, $E = ({}^-2, 1)$ and $F = ({}^-6, {}^-1)$, draw △DEF.
Reflect △DEF in the line $y = {}^-1$ and label its image D′E′F′.
Write the co-ordinates of:
(a) D′   (b) E′   (c) F′

7. If △DEF of question 6 was reflected in the line $x = 0$ (that is the y-axis) and its image labelled D″E″F‴, find the co-ordinates of:
(a) D″   (b) E″   (c) F″

8. Plot the point G(5, ${}^-7$).
(a) Reflect G in the line $x = 4$ and label the image point G′.
Write the co-ordinates of G′.
(b) Reflect G′ in the line $y = {}^-1$ and label its image G″. Write the co-ordinates of G″.

1. Copy the diagram.
   On your copy, draw the image
   when △ABC has been rotated
   90° clockwise about O. Label
   the image A′B′C′.

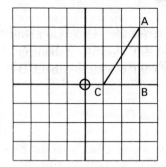

2. In the diagram, P′Q′R′ is the
   image of △PQR after it has
   been rotated 90° anticlockwise
   about O.
   Copy the diagram and draw
   and label △PQR.

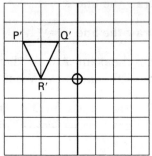

3. Draw a separate diagram for each of these rotations.
   For each one draw straight line XY but do not mark any other
   points until the point is mentioned in a question.

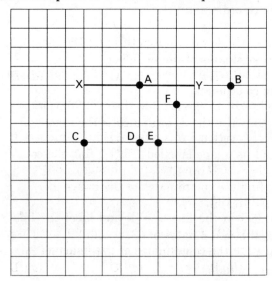

Draw the image of XY when XY is rotated through:

(a) 90° clockwise about X   (g) 180° clockwise about D

(b) 90° clockwise about A   (h) 180° anticlockwise about E

(c) 90° anticlockwise about B   (i) 90° anticlockwise about E

(d) 90° anticlockwise about C   (j) 90° clockwise about E

(e) 90° clockwise about C   (k) 270° clockwise about F

(f) 90° clockwise about D   (l) 180° clockwise about F

Exercise 8

A Use this diagram for the following questions:

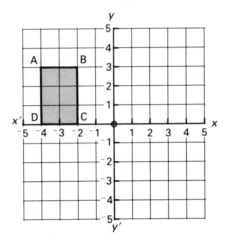

1. If rectangle ABCD was rotated through 180° about the origin O, find the co-ordinates of the image of:

(a) A   (b) B   (c) C   (d) D

2. If rectangle ABCD was rotated through 90° clockwise about the origin (0, 0), write the co-ordinates of the image of:

(a) A   (b) B   (c) C   (d) D

3. If rectangle ABCD was rotated through 180° clockwise about vertex C, find the co-ordinates of the image of:

(a) A   (b) B   (c) C   (d) D

4. If rectangle ABCD was rotated through 90° anticlockwise about vertex B, write the co-ordinates of the image of:

(a) A   (b) B   (c) C   (d) D

B Use this diagram for the following questions:

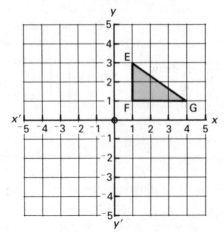

1. If triangle EFG was rotated through 180° about the origin (0, 0), find the co-ordinates of the image of:
   (a) E   (b) F   (c) G

2. If triangle EFG was rotated through 90° clockwise about the origin (0, 0), write the co-ordinates of the image of:
   (a) E   (b) F   (c) G

3. If triangle EFG was rotated 90° anticlockwise about vertex F, find the co-ordinates of the image of:
   (a) E   (b) F   (c) G

4. If triangle EFG was rotated 90° anticlockwise about vertex G, write the co-ordinates of the image of:
   (a) E   (b) F   (c) G

5. If triangle EFG was rotated 180° about vertex F, find the co-ordinates of the image of:
   (a) E   (b) F   (c) G

6. If triangle EFG was rotated through 180° about the mid-point of side EF, what would be the co-ordinates of the image of G?

7. If triangle EFG was rotated through 180° about the mid-point of side FG, what would be the co-ordinates of the image of E?

8. If triangle EFG was rotated through 180° about the mid-point of side EG, what would be the co-ordinates of the image of F?

Enlargements

1. On squared paper, draw an enlargement of the quadrilateral given, so that each side is four times as long. (That is, the scale factor is 4.)

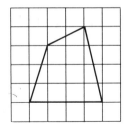

2. On squared paper, draw an enlargement of the given pentagon so that the perimeter of the enlargement is twice as big as the perimeter of the one given here.

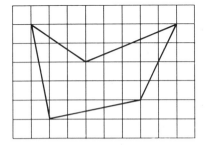

Exercise 10

1. Make a copy of △PQR. Mark a point X. Enlarge △PQR making each side twice as long. (That is, the scale factor is 2.) Use X as the centre of enlargement.

X

2. Make another copy of △PQR. Mark point Y. Enlarge △PQR using a scale factor of 2, where Y is the centre of enlargement.

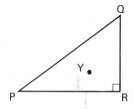

3. Make another copy of △PQR. Enlarge △PQR making each side twice as long, where P is the centre of enlargement.

4. Construct triangle ABC where AB = 30 mm, AC = 25 mm and BC = 15 mm. Mark a point P where PA = 30 mm and PC = 45 mm (P should be near the left-hand edge of your page).

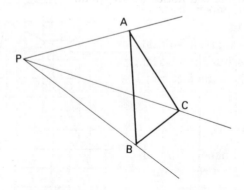

Enlarge △ABC using P as the centre of enlargement and a scale factor of 3.

Exercise 11

1. Draw a pair of axes as shown opposite.

Plot the points A(1, 4), B(4, 6) and C(5, 2).
Join the points to form △ABC.
Enlarge △ABC using the origin (0, 0) as the centre of enlargement and a scale factor of 3.
Label the image A′B′C′.
Find the ratios:

$$\frac{A'B'}{AB}, \quad \frac{B'C'}{BC} \quad \text{and} \quad \frac{C'A'}{CA}$$

then write what you notice about them.

2. Draw another pair of axes as shown opposite.
Plot the points P(2, 6), Q(6, 8), R(8, 4) and S(6, 1). Join PQRS to form a quadrilateral. Enlarge PQRS using (0, 0) as the centre of enlargement and a scale factor of 2. Label the image P′Q′R′S′.

Find the ratios:

$$\frac{P'Q'}{PQ}, \quad \frac{Q'R'}{QR}, \quad \frac{R'S'}{RS} \quad \text{and} \quad \frac{S'P'}{SP}$$

then write what you notice about them.

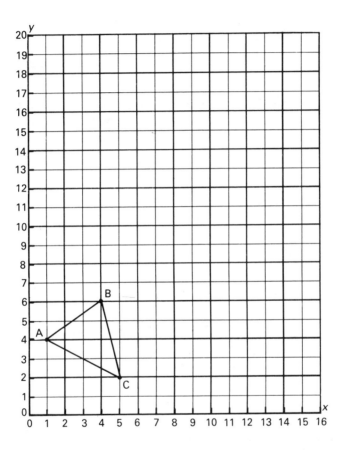

A **1.** In question 1 of Exercise 11 compare the co-ordinates of A and A′, B and B′, C and C′, then write what you notice.

2. In question 2 of Exercise 11 compare the co-ordinates of P and P′, Q and Q′, R and R′, S and S′, then write what you notice.

B **1.** Draw a pair of axes as shown where the *x*-values range from 0 to 14 and the *y*-values from 0 to 15.

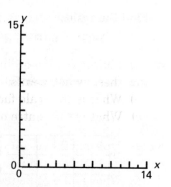

2. Plot the points E(3, 3), F(6, 3), G(2, 7) then join them to form a triangle.

3. Multiply each of the co-ordinates in question 2 by 2 then plot and join these new points. Label the three new points E′, F′ and G′, where the co-ordinates are double those of E, F and G respectively.

4. Find the ratios:

(a) $\dfrac{E'F'}{EF}$ (b) $\dfrac{F'G'}{FG}$ (c) $\dfrac{G'E'}{GE}$

5. Write what you notice about the ratios.

6. △E′F′G′ is an enlargement of △EFG. What is the scale factor of the enlargement?

Note The ratio of corresponding sides of the image to the object gives the scale factor of the enlargement.

C **1.** Draw a pair of axes as shown where the *x*-values range from ⁻8 to ⁺8 and the *y*-values range from ⁻10 to ⁺10.

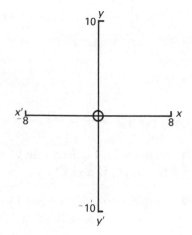

2. Plot the points P(0, 2), Q(2, ⁻2) and R(⁻1, ⁻2). Join them to form a triangle.

3. (a) Multiply the co-ordinates of P, Q and R by 4 then plot and join these new points.

(b) What is the scale factor of the enlargement that is obtained?

(c) What is the centre of the enlargement?

4. (a) Multiply the co-ordinates of P, Q and R by 3 then plot and join these image points.

(b) What is the scale factor of this enlargement?

(c) What is the centre of enlargement?

5. (a) Multiply the co-ordinates of P, Q and R by 5 then plot and join these image points.

(b) What is the scale factor of this enlargement?

(c) What is the centre of enlargement?

D For these questions, find the co-ordinates of the images of the vertices of the shapes given. The centre of enlargement is $(0, 0)$.

1. $J(1, 3)$, $K(4, 1)$, $L(2, 1)$ is a triangle. The scale factor = 2.

2. $J(1, 3)$, $K(4, 1)$, $L(2, 1)$ is a triangle. The scale factor = 5.

3. $M(5, 5)$, $N(1, 4)$, $O(0, 0)$, $P(6, 0)$ is a quadrilateral. The scale factor = 6.

4. $Q(0, 3)$, $R(3, 5)$, $S(4, 2)$, $T(1, 0.5)$ is a quadrilateral. The scale factor = 4.

Exercise 13   M

A Copy the following. For each question find the centre of enlargement.

1.

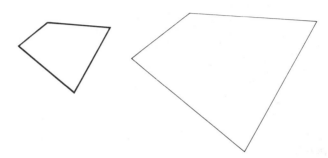

2.

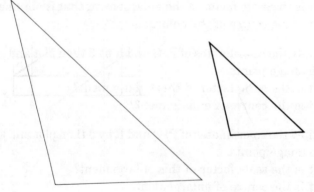

3.   **4.**

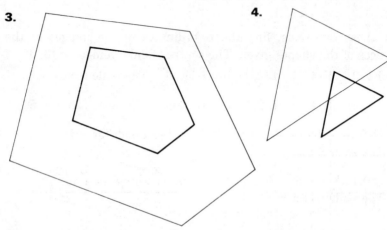

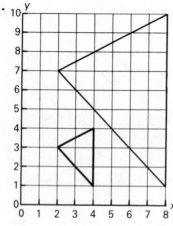

B Copy on to squared paper then find the co-ordinates of the centre of enlargement:

1.

2.

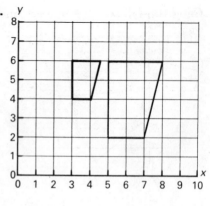

236

3.

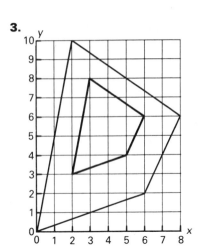

6.

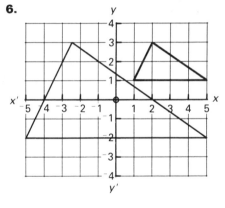

4.

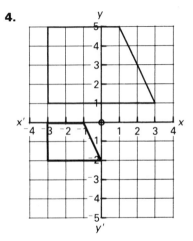

7.

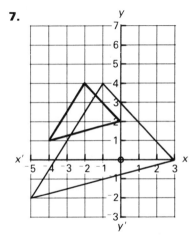

5.

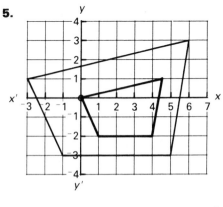

8.

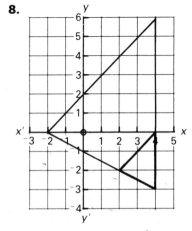

237

C 1. For each question in part A, write the scale factor of the enlargement.

2. For each question in part B, write the scale factor of the enlargement.

Exercise 14

A Copy the diagrams given below. For each question, enlarge △PQR using the scale factor given. C is the centre of enlargement.

1. Scale factor = 2

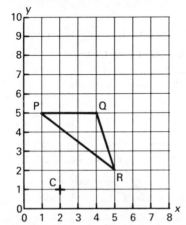

2. Scale factor = 3

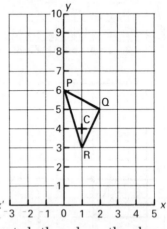

B For each question, draw axes as instructed, then draw the shape given. In each case, draw the image of the shape under an enlargement, using the scale factor and centre of enlargement as stated.

1. The x-axis is labelled from 0 to 8 and the y-axis from 0 to 10.
D(3, 6), E(1, 5), F(5, 4) is a triangle.
Scale factor is 2. Centre of enlargement is C(2, 2).

2. The x-values range from 0 to 8 and the y-values range from 0 to 10.
G(1, 3), H(3, 3), I(3, 1) is a triangle.
Scale factor is 3. Centre of enlargement is C(1, 0).

3. The x-values range from 0 to 8 and the y-values range from 0 to 10.
J(5, 6), K(2, 3), L(6, 2) is a triangle.
Scale factor is 2. Centre of enlargement is C(4, 3).

238

4. The *x*-values range from ⁻4 to ⁺4 while the *y*-values range from ⁻5 to ⁺5.

M(1, 2), N(⁻3, 3), P(⁻3, ⁻2) is a triangle.

Scale factor is 2. Centre of enlargement is C(⁻2, 1).

5. The *x*-values range from 0 to 8 while the *y*-values range from ⁻4 to ⁺6.

Q(0, 2), R(1, 5), S(2, 3) is a triangle.

Scale factor is 3. Centre of enlargement is C(0, 5).

6. The *x*-values range from ⁻3 to ⁺5 and the *y*-values range from ⁻5 to ⁺5.

T(0, 2), U(2, 3), V(3, ⁻1), W(⁻1, ⁻1) is a quadrilateral.

Scale factor is 2. Centre of enlargement is C(1, 1).

7. The *x*-values range from ⁻4 to ⁺4 while the *y*-values range from ⁻6 to ⁺4.

X(4, ⁻5), Y(3, ⁻2), Z(1, ⁻1), A(0, ⁻4) is a quadrilateral.

Scale factor is 2. Centre of enlargement is C(4, ⁻6).

8. The *x*-values range from ⁻3 to ⁺5 while the *y*-values range from ⁻5 to ⁺5.

B(1, 1), C(2, 4), D(3, 2) is a triangle.

Scale factor is 3. Centre of enlargement is C(2, 4).

Exercise 15   Non-Integral Scale Factors

Find, for example, the scale factor of this enlargement:

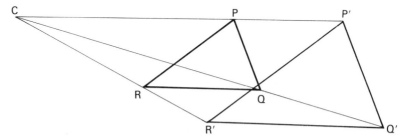

$$\frac{CP'}{CP} = \frac{CQ'}{CQ} = \frac{CR'}{CR} = \frac{3}{2}. \text{ Also } \frac{P'Q'}{PQ} = \frac{Q'R'}{QR} = \frac{R'P'}{RP} = \frac{3}{2}.$$

Since all lengths on the enlargement are $1\frac{1}{2}$ times as big as the corresponding lengths on the object, the scale factor $= 1\frac{1}{2}$ (1.5).

Find the scale factor of these enlargements:

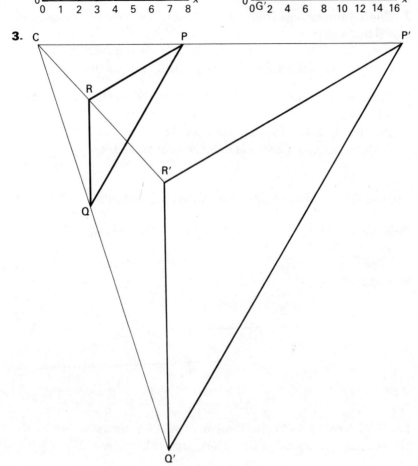

Exercise 16   Miscellaneous Transformations

Consider these statements for transformations:

A   All points on the object move the same distance.

B   The image and the object have the same shape and size.

C   Transformations exist in which only one point on the object does not move.

D   Transformations exist in which more than one point on the object do not move.

E   All points on the object move in the same direction.

Write which of the statements above are true for:

1. translations
2 reflections
3. rotations
4. enlargements

19 Ratio, Proportion and Speed

Ratio

Exercise 1

1. A certain sum of money was shared between Simon and Trevor so that Simon received £12 and Trevor £30. Write these two amounts as a ratio of the smallest amount to the largest amount in the following ways.
   (a) as a ratio of whole numbers in its simplest form
   (b) in the form $1 : m$
   (c) in the form $n : 1$

2. The ratio height : span gives the pitch of a roof of a house. This can be written as:

   $$\text{Pitch} = \frac{\text{height}}{\text{span}}$$

   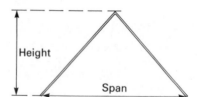

   (a) Calculate the pitch of each roof for the buildings given in the following table. (Use the form $n : 1$.)

Building	Height (m)	Span (m)	Pitch
A	3.6	10	
B	3.15	9	
C	3.4	8.5	

   (b) Which of the buildings in part (a) has the steepest roof?

3. Here is a set of numbers: $\{1, 4, 6, 12, 16, 18, 24\}$.
   Find, in its simplest form, the ratio of the multiple of 9 to the largest square number in the set.

242

4. The sides of a triangle measure 8.1 cm, 6.3 cm and 5.4 cm. Find, in its simplest form, the ratio of:

(a) the shortest side to the longest side,

(b) the longest side to the middle-sized side,

(c) the middle-sized side to the shortest side.

5. A mother was 24 years old when her son was born. Consider the ratio of the ages of the mother and her son.

(a) Find the ratio when the mother is 28 years old. (Give the ratio in its simplest form.)

(b) Find, in its simplest form, the ratio of their ages when the son is 15 years old.

(c) Find their ages when the ratio is 2 : 1.

(d) Find their ages when the ratio is 5 : 1.

(e) Find their ages when the ratio is 3 : 1.

(f) Find their ages when the ratio is 3 : 2.

(g) Does the ratio of their ages get bigger or smaller as they get older?

Exercise 2

(Throughout this exercise, find the ratio of the larger shape to the smaller and write each answer in its simplest form.)

1. The sides of the two squares measure 15 mm and 20 mm. Find the ratio of:

(a) the lengths of their sides,

(b) their perimeters.

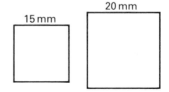

2. The sides of two equilateral triangles measure 4.8 cm and 3.2 cm. Find the ratio of:

(a) the lengths of their sides,

(b) their perimeters.

3. Two circles have diameters of 24 mm and 18 mm. Find the ratio of:

(a) their diameters,

(b) their radii,

(c) their circumferences.

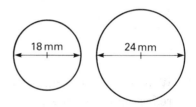

4. A rectangle measures 4.2 cm by 3.5 cm. Its enlargement has sides that are three times as long. Find the ratio of:

(*a*) the lengths,   (*b*) the breadths,   (*c*) the perimeters.

The gear wheels shown are joined by a chain. If the large wheel has 40 teeth and the small wheel 20 teeth, when the large wheel makes 1 full revolution the chain moves 40 links and the small wheel turns by 40 teeth which is 2 revolutions.

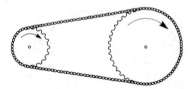

When two gear wheels fit together as shown where the teeth of one are turned by the teeth of the other, they are said to be *in mesh*. If the large wheel has 40 teeth and the small wheel 20 teeth, when the large wheel makes 1 full revolution its 40 teeth move the small wheel by 40 teeth which is 2 full turns. This is the same as for the chain-driven wheels above.

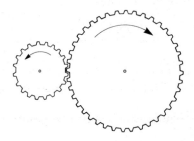

Note   The smaller the number of teeth, the greater the number of revs. The greater the number of teeth, the smaller the number of revs.

If a wheel makes more revs during a certain period of time, it must be turning faster. Therefore, the smaller the number of teeth the faster the wheel turns; the greater the number of teeth the slower the wheel turns.

If the turning of the large wheel turns the small wheel then the large wheel is called the *driver* and the small wheel the *follower*.

We can write the formula:

$$\frac{\text{Speed of follower}}{\text{Speed of driver}} = \frac{\text{number of teeth on driver}}{\text{number of teeth on follower}}$$

Some wheels are belt-driven. (They are usually called pulleys.) They have no teeth.

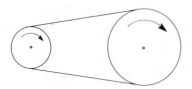

For belt-driven wheels:

$$\frac{\text{Speed of follower}}{\text{Speed of driver}} = \frac{\text{diameter of driver}}{\text{diameter of follower}}$$

(*Note*

Ratio of the circumferences = ratio of the diameters.)

If we refer to the diameter of a wheel or the number of teeth on a wheel as its *size*, the formula becomes:

$$\frac{\text{Speed of follower}}{\text{Speed of driver}} = \frac{\text{size of driver}}{\text{size of follower}}$$

If the driver is larger than the follower then its speed must be slower than that of the follower. Then:

$$\frac{\text{Speed of follower}}{\text{Speed of driver}} \text{ must work out to be greater than 1}$$

and

$$\frac{\text{Size of driver}}{\text{Size of follower}} \text{ must also be greater than 1}$$

The last two ratios can help in the writing of the formula.

Note For the chain-driven and belt-driven wheels, both wheels turn in the same direction. However, for wheels in mesh, when one wheel turns clockwise the other turns anti-clockwise.

Exercise 3

A **1.** A gear wheel with 36 teeth drives another wheel with 20 teeth.
   (*a*) If the speed of the driver was 50 rev/min, find the speed of the follower.
   (*b*) If the speed of the follower was 72 rev/min, find the speed of the driver.

2. A pulley with a diameter of 30 cm revolves at 160 rev/min. It drives a 24 cm pulley. At how many rev/min does the 24 cm pulley turn?

3. An electric motor turns a 4 cm pulley at 1800 rev/min. This drives a 30 cm pulley. What is the speed of the driven pulley in rev/min?

4. A gear wheel with 56 teeth and turning at 30 rev/min drives another gear wheel at 80 rev/min. How many teeth has this follower gear wheel?

B   Three gear wheels are in mesh as shown.
   The large wheel has 60 teeth, the centre wheel has 40 teeth and the small wheel has 30 teeth.

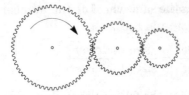

1. If the large wheel turns clockwise, in which direction will the small wheel turn?

2. If the large wheel turns at 36 r.p.m. (rev/min), find the r.p.m. at which the other wheels turn.

3. If the middle wheel were removed, the large wheel could be placed directly in mesh with the smallest wheel. If the large wheel were still set to turn at 36 r.p.m., at what rate would the small wheel turn?

4. Write what you notice about the speed of the small wheel in your answers to questions 2 and 3.

Scales

Exercise 4

e.g. A scale of $1 : 20\,000$ is the same as $1\,\text{cm}$ to $20\,000\,\text{cm}$,
that is, $1\,\text{cm}$ to $200\,\text{m}$.

A For each scale given as a ratio, rewrite it using the required units:

1. $1 : 400$ is the same as $1\,\text{cm}$ to $\boxed{?}$ m.
2. $1 : 40\,000$ is the same as $1\,\text{cm}$ to $\boxed{?}$ m.
3. $1 : 200\,000$ is the same as $1\,\text{cm}$ to $\boxed{?}$ km.
4. $1 : 500\,000$ is the same as $1\,\text{cm}$ to $\boxed{?}$ km.
5. $1 : 1\,000\,000$ is the same as $1\,\text{cm}$ to $\boxed{?}$ km.
6. $1 : 6\,000\,000$ is the same as $1\,\text{cm}$ to $\boxed{?}$ km.
7. $1 : 250\,000$ is the same as $1\,\text{cm}$ to $\boxed{?}$ km.
8. $1 : 75\,000$ is the same as $1\,\text{cm}$ to $\boxed{?}$ m.

B Write each of the following scales in the form $1 : n$:

1. $1\,\text{cm}$ to $5\,\text{m}$ is the same as $1 : \boxed{?}$.
2. $1\,\text{cm}$ to $30\,\text{m}$ is the same as $1 : \boxed{?}$.
3. $1\,\text{cm}$ to $700\,\text{m}$ is the same as $1 : \boxed{?}$.
4. $1\,\text{cm}$ to $250\,\text{m}$ is the same as $1 : \boxed{?}$.
5. $1\,\text{cm}$ to $2\,\text{km}$ is the same as $1 : \boxed{?}$.
6. $1\,\text{cm}$ to $50\,\text{km}$ is the same as $1 : \boxed{?}$.
7. $1\,\text{cm}$ to $80\,\text{km}$ is the same as $1 : \boxed{?}$.
8. $1\,\text{cm}$ to $7.5\,\text{km}$ is the same as $1 : \boxed{?}$.

Exercise 5

1. A model is made to a scale of $1 : 20$.
   (*a*) Find its actual length if the model is $8\,\text{cm}$ long.
   (*b*) Find its actual length if the model is $9.4\,\text{cm}$ long.
   (*c*) Find its actual width if the model is $6\,\text{cm}$ wide.

(*d*) Find the model's width if the actual width is 80 cm.

(*e*) Find the model's length if the actual length is 3 m.

(*f*) Find the model's length if the actual length is 1.4 m.

2. A plan of a house has a scale of 1 : 50.

(*a*) On the plan, a room is 10 cm long and 8 cm wide. What are its true dimensions?

(*b*) If a room is 11 cm long on the plan, what is its true length?

(*c*) A room is 6 m long and 3.5 m wide. What would these dimensions measure on the plan?

3. A map is drawn to a scale of 1 : 50 000.

(*a*) What distance is represented by 1 cm?

(*b*) What distance is represented by 1 mm?

(*c*) Two places are 6 cm apart on the map. What is the actual distance between the places?

(*d*) Two places are 4.8 cm apart on the map. What is the actual distance between the places?

(*e*) Two places are 4 km apart. What is the map distance?

(*f*) Two places are 2.8 km apart. What is the map distance?

Average Speed

$$\text{Average speed} = \frac{\text{total distance travelled}}{\text{total time taken}}$$

$$\text{Distance travelled} = \text{average speed} \times \text{time taken}$$

$$\text{Time taken} = \frac{\text{distance travelled}}{\text{average speed}}$$

If distance is in kilometres and time in hours then average speed is in kilometres per hour (km/h).

If distance is in miles and time in hours then average speed is in miles per hour (m.p.h.).

If distance is in metres and time in seconds then average speed is in metres per second (m/s or $m\,s^{-1}$).

248

Copy and complete the tables:

A

	Distance travelled (km)	Time taken	Average speed (km/h)
1.	150	3 h	
2.	320	4 h	
3.	130	2 h	
4.	219	3 h	
5.	574	6 h	
6.		3 h	70
7.		2 h	89
8.		4 h	112
9.	180		60
10.	375		75

	Distance travelled (km)	Time taken	Average speed (km/h)
11.	210		60
12.		$1\frac{1}{2}$ h	40
13.		$2\frac{1}{4}$ h	80
14.	250	2 h 30 min	
15.	150		120
16.		$1\frac{3}{4}$ h	64
17.	189		81
18.	120	1 h 15 min	
19.	228	$\frac{1}{2}$ h	
20.		$1\frac{1}{2}$ h	6

B

	Distance travelled (km)	Starting time	Finishing time	Time taken	Average speed (km/h)
1.	270	13.45	16.45		
2.		19.20	21.20		70
3.	250	08.35			50
4.	100	11.55	14.25		
5.		12.08		6 h	72
6.	340	07.15			85
7.			11.05	$2\frac{1}{2}$ h	20
8.	1470	10.49		3 h 30 min	
9.		06.32	08.12		360
10.	315	09.50	13.35		

Exercise 7

1. Andrea walked 18 km in 3 h. Find her average speed of walking.

2. A car travelled at an average speed of 63 km/h on a journey that took 4 h. How far was the journey?

3. An aeroplane travelled at 450 km/h for $4\frac{1}{2}$ h. How far did it travel?

4. A train averaged 92 km/h on a journey of 460 km. How long did the journey take?

5. A train travelled 288 km in 4 h while a car travelled 355 km in 5 h. Which travelled at the faster average speed?

6. At an average speed of 56 km/h, a bus takes 5 h to travel a certain journey. How long would the bus take if it averaged 70 km/h?

7. (a) A car travelled 180 km in 3 h. What was its average speed?
   (b) If the car in part (a) then travelled 140 km during the next 2 h, what was its average speed for that part of the journey?
   (c) Find the car's average speed for the full journey of 320 km (from parts (a) and (b)).

8. If Kate walked 8 km in 1 h then walked at 5 km/h for the next 2 h:
   (a) What would be the total distance travelled?
   (b) What would be the total time of the walk?
   (c) What would be Kate's average speed for the walk?

9. Travelling by car, if I can average 60 km/h, at what time must I set off on a 270 km journey to arrive at my destination by 14.15?

10. Two cars race each other over a total distance of 750 km. One car averages 150 km/h while the other averages 125 km/h:
    (a) By how many kilometres will the faster car win?
    (b) How many times will the slower car be lapped if each lap is 30 km?

Exercise 8

1. Guy walked 12 miles in 4 h. Find his average speed of walking.

2. A bus travelled 228 miles in 6 h. Find its average speed.

3. A lorry travelled at a steady speed of 34 m.p.h. How far did it travel in:
   (a) 2 h? (b) 5 h? (c) $3\frac{1}{2}$ h?

4. A train travelled at a steady speed of 60 m.p.h. How long did the following journeys take?
   (a) 420 miles (b) 150 miles (c) 220 miles

5. A car travels for $2\frac{1}{2}$ h at 44 m.p.h. while a bus travels for 3 h at 38 m.p.h. Which travels the further and by how many miles?

6. If an aeroplane averages 320 m.p.h., how far will it travel in:
   (a) 7 h? (b) $4\frac{1}{2}$ h? (c) 3 h 15 min?

7. A cyclist travels at 9 m.p.h. for 6 h. How long will it take a car to travel that same distance at 36 m.p.h?

8. (a) Two cars set off together on a journey of 216 miles. If one averaged 54 m.p.h. and the other 48 m.p.h., how much earlier does the faster car arrive at the destination?
   (b) If the journey in part (a) was only 144 miles, how much earlier would the faster car arrive at the destination?

Exercise 9   Direct Proportion

1. The table gives sets of numbers that are in *direct proportion*. It gives suggested quantities of wood ash needed to dress certain areas of soil when used as a fertiliser. Copy and complete the table.

Quantity of wood ash (kg)	1	2	3		7			15	20	
Area covered (m²)			12	16		36	48			100

2. The numbers in these tables are in direct proportion. However, in each table, two of the numbers have been wrongly entered. Find the wrong entries and write what the correct entries should be.

(a)

3	5	6	8	9	10	12	13	15	16
24	40	48	64	76	80	96	104	120	124

(b)

2	3	4	6	8	10	12	15	20	25
3	$4\frac{1}{2}$	6	8	12	15	16	$22\frac{1}{2}$	30	$37\frac{1}{2}$

3. One book needs 2 cm of shelf space. How much space do 7 books need if all the books are identical in size?

4. If my pulse beats 72 times in 1 min, how many times will it beat in 3 min if it continues to beat regularly?

5. If $2\,m^3$ of sand has a mass of 5.2 t, what would be the mass of $6\,m^3$ of sand?

6. At a steady rate I walked 15 km in 3 h. How far did I walk in:
(a) 1 h?   (b) 5 h?   (c) 4 h?

7. An electric fire uses 12 units of electricity in 6 h:
(a) How many units does it use in 3 h?
(b) How many units does it use in 1 h?
(c) How many units does it use in 4 h?
(d) How many hours does it take to use 24 units?
(e) How many hours does it take to use 4 units?
(f) How many hours does it take to use 10 units?

8. Repeat question 7 for an electric fire that uses 9 units of electricity in 6 h.

9. A car travels 136 miles on 4 gal of petrol. How far will it travel on:
(a) 2 gal?   (b) 3 gal?

10. A car travelled 76 miles in 2 h. At the same steady speed, how far will it travel in:
(a) 3 h?   (b) 5 h?

11. 8 cm³ of gold weights 154.4 g. Find the mass of:

    (a) 4 cm³   (c) 6 cm³   (e) 0.8 cm³   (g) 12.5 cm³

    (b) 16 cm³   (d) 10 cm³   (f) 0.7 cm³   (h) 30 cm³

12. In a hydraulic press a force of 80 N will lift 32 000 N.

    (a) What will 40 N lift?

    (b) What will 60 N lift?

    (c) What force will lift 8000 N?

    (d) What force will lift 20 000 N?

13. The resistance of a wire is directly proportional to its length. If 3 m of wire has a resistance of 11.25 Ω what would be the resistance of lengths of:

    (a) 2 m?   (b) 5 m?   (c) 8 m?

14. 5 cm on a map represents 3 km:

    (a) What distance is represented by 3 cm?

    (b) What distance is represented by 14 cm?

    (c) What distance is represented by 7.5 cm?

    (d) What length represents a distance of 12 km?

    (e) What length represents a distance of 7.5 km?

15. A scale model of a car is 6 cm long and the real car is 4 m long. The width of the real car is 1.6 m. What is the width of the model car?

20 **Similarity**

Two figures are *similar* if they have the *same shape*.

Exercise 1

These two figures are similar.
(They do not have to face the
same direction.)

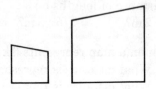

For each question, write whether or not the two given figures are similar:

1.

2.

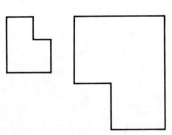

3.

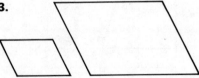

4.

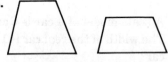

5.

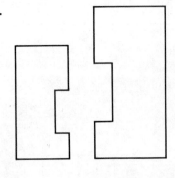

6.

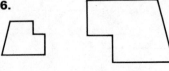

Exercise 2

For each question, write whether or not the two given shapes are similar:

1.

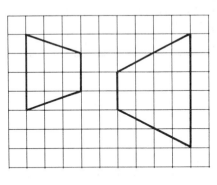

4.

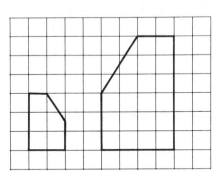

2.

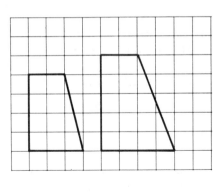

5.

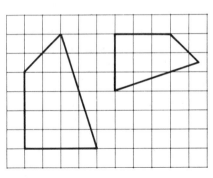

3.

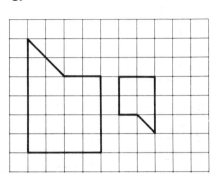

6.

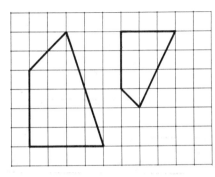

7.

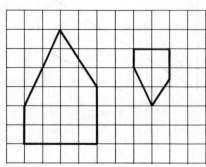

8.

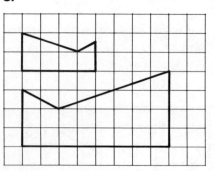

Similar Triangles

Two triangles are similar if their angles are equal.

Exercise 3

In each question, write whether the two triangles are similar (if only two angles are given you may need to calculate the third):

1.

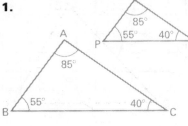

3.

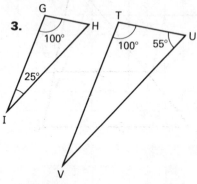

2.

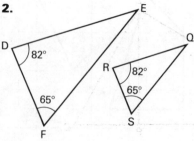

4.

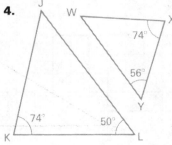

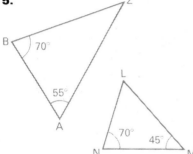

6.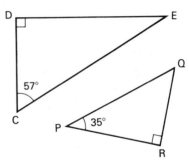

Exercise 4

Copy the given diagrams. On each one, mark the angles that are equal and name the similar triangles.

1.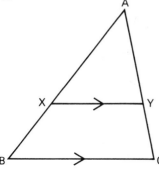

2.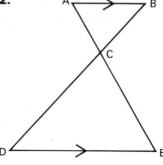

3.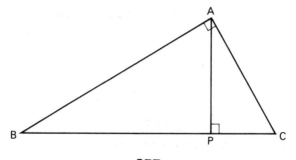

Exercise 5

A The two triangles in the diagram are similar:

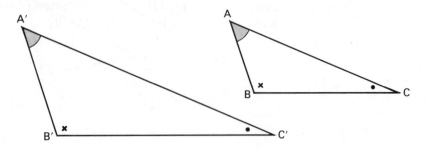

1. By measuring then calculating, find the ratios $\dfrac{A'B'}{AB}$, $\dfrac{B'C'}{BC}$ and $\dfrac{C'A'}{CA}$ giving them in the form $m : 1$.

2. Write what you notice about the three ratios in question 1.

B Draw some pairs of similar triangles of your own.
Label them ABC and A'B'C'.
For each pair of triangles find:

$$\dfrac{A'B'}{AB}, \quad \dfrac{B'C'}{BC}, \quad \text{and} \quad \dfrac{C'A'}{CA} \quad \text{in the form} \quad m : 1.$$

Write what you notice about these ratios.

C Draw any triangle and label it ABC.
Draw a line XY parallel to BC.
Measure necessary lengths, then calculate the following ratios giving answers in the form $m : 1$.

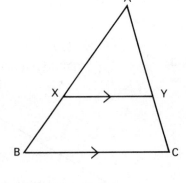

1. (a) $\dfrac{AB}{AX}$ (b) $\dfrac{AC}{AY}$ (c) $\dfrac{BC}{XY}$

2. (a) $\dfrac{AX}{AB}$ (b) $\dfrac{AY}{AC}$ (c) $\dfrac{XY}{BC}$

3. Write what you notice about these ratios.

D Draw any pair of parallel lines as shown.
Label them PQ and ST.
Join Q to S and P to T.
Let these lines cross at X.

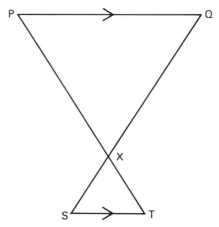

Measure necessary lengths then calculate the following ratios giving answers in the form $m : 1$.

1. (a) $\dfrac{PQ}{TS}$ (b) $\dfrac{PX}{TX}$ (c) $\dfrac{QX}{SX}$

2. (a) $\dfrac{TS}{PQ}$ (b) $\dfrac{TX}{PX}$ (c) $\dfrac{SX}{QX}$

3. Write what you notice about these ratios.

If two triangles are similar then their corresponding sides are in the same ratio.

The triangles shown are similar. ($\hat{A} = \hat{R}$, $\hat{B} = \hat{Q}$ and $\hat{C} = \hat{P}$.)

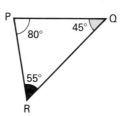

We can write:
△ABC and △RQP are similar.

Note the angle order:

Since A is given first,	R must be given first	(since $\hat{A} = \hat{R}$).
Since B is given second,	Q must be given second	(since $\hat{B} = \hat{Q}$).
Since C is given third,	P must be given third	(since $\hat{C} = \hat{P}$).

e.g. In the diagram, △XYZ and △NML are similar (the angles are equal). We can therefore write the ratios:

$$\frac{XY}{NM} = \frac{YZ}{ML} = \frac{ZX}{LN}$$

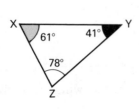
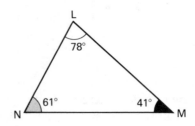

For each pair of similar triangles in Exercise 3 (p.256–7) write three equivalent ratios as in the example above.

> If the corresponding sides of two triangles are in the same ratio then the two triangles are similar.

Exercise 7

A For each question, find whether the two triangles given are similar:

1.

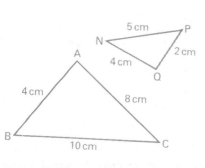

2.

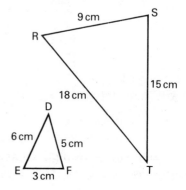

3.

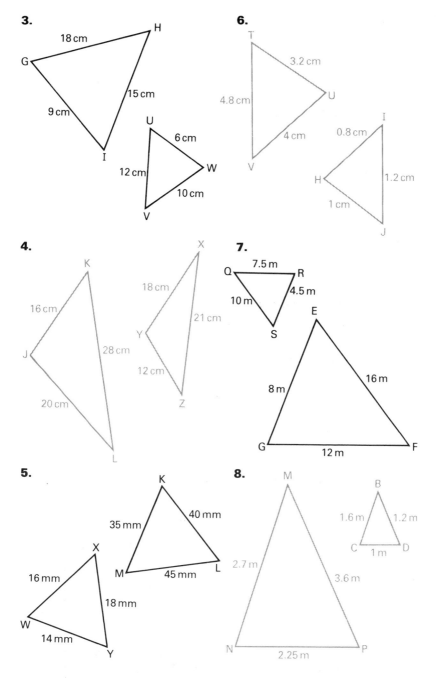

18 cm
G
H
15 cm
9 cm
I

U
6 cm
12 cm
W
10 cm
V

6.

T
3.2 cm
4.8 cm
U
4 cm
V

I
0.8 cm
H
1.2 cm
1 cm
J

4.

K
16 cm
J
28 cm
20 cm
L

X
18 cm
21 cm
Y
12 cm
Z

7.

7.5 m
Q
R
4.5 m
10 m
S

E
8 m
16 m
G
12 m
F

5.

K
35 mm
40 mm
M
45 mm
L

X
16 mm
18 mm
W
14 mm
Y

8.

M
2.7 m
3.6 m
N
2.25 m
P

B
1.6 m
1.2 m
C
1 m
D

B For each pair of triangles that are similar in part A, write which angles are equal.

261

C By measuring sides then calculating ratios find whether or not the triangles below are similar:

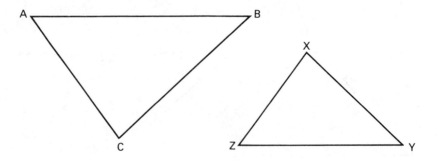

Exercise 8

Two sides of a triangle measure 4 cm and 3 cm and one angle measures 40°. In another triangle, two sides measure 8 cm and 6 cm (they are twice as long) while one angle measures 40°.

Must the two triangles be similar?—the questions in this exercise should help you to answer this.

1. Are triangles ABC and PQR similar?
(They have been drawn to the correct size.)

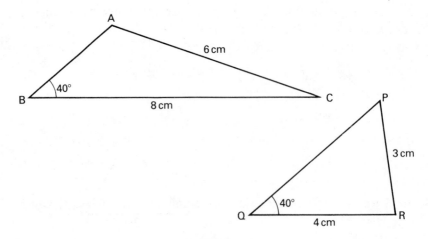

2. Are triangles ABC and PQR similar?

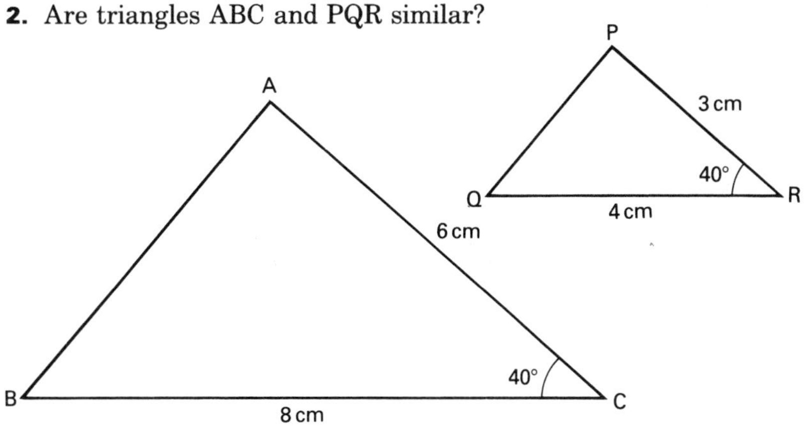

The same information was used in both questions of Exercise 8. However, the triangles in question 1 are not similar while those in question 2 are similar.

The difference is in the position of the given angle.

The triangles are similar if the given angle is the *included angle* (that is, it lies between the two given sides).

> Two triangles are similar if two pairs of sides are in the same ratio *and* the angles between those pairs of sides are equal.

Exercise 9

For each question, find whether the two triangles given are similar. (The marked angles in each question are equal.)

1.

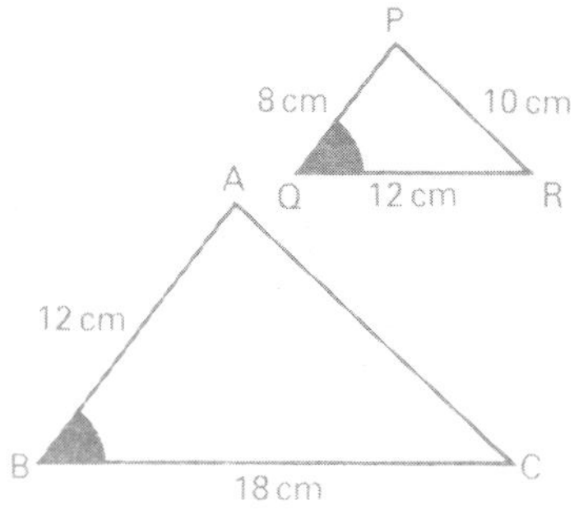

2.

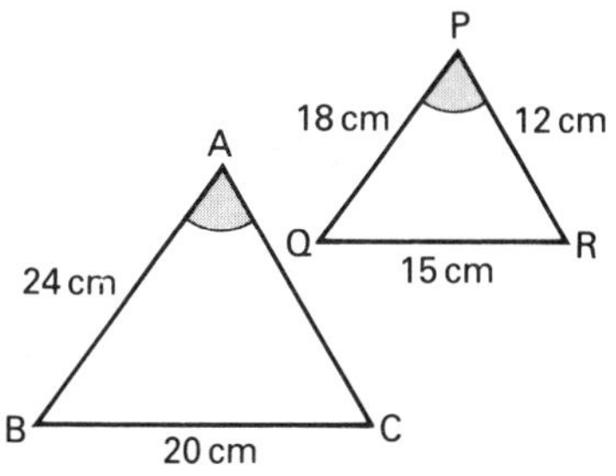

263

3.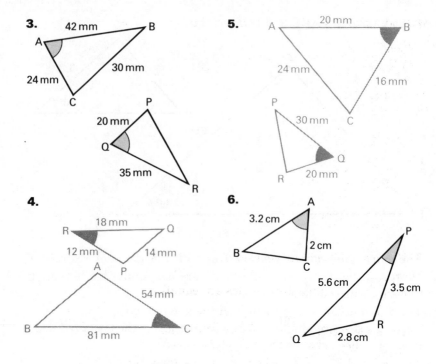

5.

4.

6.

7. In △ABC and △PQR: Â = P̂, AB = 16 cm, AC = 24 cm, BC = 20 cm, PQ = 10 cm and PR = 15 cm.

8. In △ABC and △PQR: ∠B = ∠R, AB = 20 cm, AC = 30 cm, BC = 40 cm, PQ = 14 cm and QR = 28 cm.

The ideas of ratio and of scale factors can be used to find sides of similar triangles.

Here are two methods of finding the missing side PQ of △PQR shown.

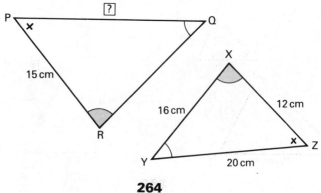

Method 1 (Ratio method)

$\triangle \substack{PQR \\ ZYX}$ are similar   (the angles are equal)

so $\dfrac{PQ}{ZY} = \dfrac{PR}{ZX}$

∴ $\dfrac{PQ}{20} = \dfrac{15}{12}$

∴ $PQ = \dfrac{15}{12} \times 20$

∴ $\underline{PQ = 25\,cm}$

Method 2   (Scale factor method)

$\triangle \substack{PQR \\ ZYX}$ are similar (the angles are equal)

The scale factor $= \dfrac{PR}{ZX} = \dfrac{15}{12} = \dfrac{5}{4} = 1.25.$

∴ $PQ = 1.25 \times ZY$ (Since PQ is the image of ZY.)

so $PQ = (1.25 \times 20)\,cm$

∴ $\underline{PQ = 25\,cm}$

Exercise 10

Calculate the required sides:

A **1.**   **2.**

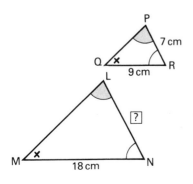

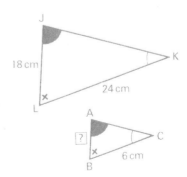

265

3.

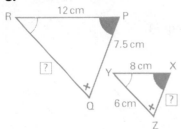

5.

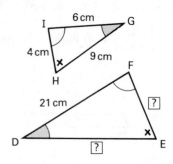

4.

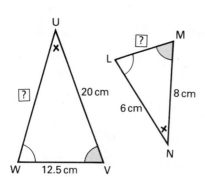

6.

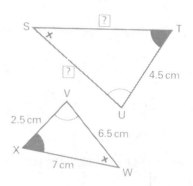

B **1.** △GHI and △RST are similar. Ĝ = R̂ and Ĥ = Ŝ.
HI = 12 cm, GI = 16 cm, RS = 21 cm, RT = 24 cm. Find:
(a) ST (b) GH

2. △DEF and △TUV are similar. ∠E = ∠U and ∠F = ∠V.
DE = 9 cm, EF = 4.5 cm, TV = 12 cm and TU = 15 cm. Find:
(a) UV (b) DF

3. In △FGH and △RST, angle F = angle R and angle
G = angle T. GH = 36 cm, FH = 27 cm, RT = 10 cm and
ST = 16 cm. Find:
(a) RS (b) FG

4. In △KLM and △NPQ, ∠K = ∠P and ∠M = ∠N. KL = 4 m,
KM = 3.2 m, PQ = 1.5 m and NQ = 2.1 m. Find:
(a) LM (b) PN

C **1.**

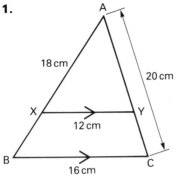

AX = 18 cm, AC = 20 cm,
XY = 12 cm, BC = 16 cm.
Find:
(a) AY (b) AB (c) XB

3.

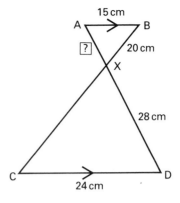

AB = 15 cm, BX = 20 cm,
DX = 28 cm, CD = 24 cm.
Find: (a) CX (b) AX

2.

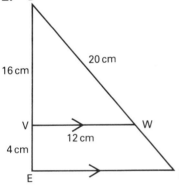

DW = 20 cm, DV = 16 cm,
VE = 4 cm, VW = 12 cm.
Find:
(a) EF (b) DF (c) WF

4.

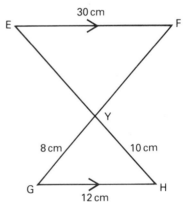

EF = 30 cm, YG = 8 cm,
GH = 12 cm, HY = 10 cm.
Find EY.

D For extra practice, find the unknown sides in the questions in
Exercise 9 where the triangles are similar.

1. How tall is the poplar tree?

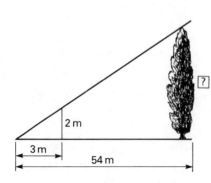

4. How far is the boat from the shore?

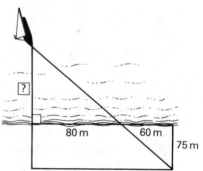

2. How high is the flag-pole?

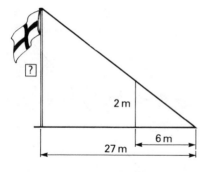

5. How far is the house from the road?

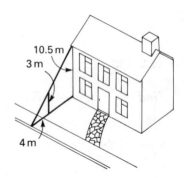

3. How wide is the river?

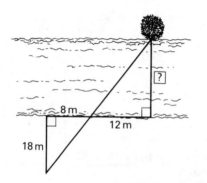

6. How high is the church spire?

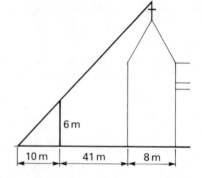

7. How far is the boat from the shore?

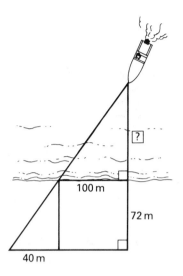

Exercise 12

To Find the Height of a Building (or other tall object)

1. Cut out of card, two right-angled triangles as shown and stick a straw along each hypotenuse.

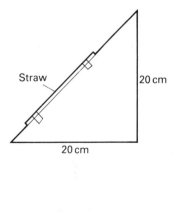

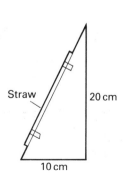

2. (a) Position yourself at a distance from the building so that it is possible to sight the top of the building using one of the cardboard triangles (as shown in the sketch).

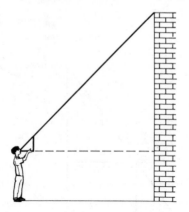

(b) Measure the distance from the building to the point from which the top of the building can be sighted.
(c) Measure your own height from the ground up to eye level.
(d) Find the height of the building by calculation. (Use similar triangles.)

3. Repeat question 2 using the other cardboard triangle.

Exercise 13   Other Similar Figures

1. Calculate the distance from North-East Point to the mouth of the River Westworth.
(The scale factor is 100 000.)

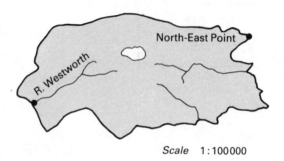

Scale   1 : 100 000

2. How far is it from South-West View to Pirates' Bay.
(The scale factor is 250 000.)

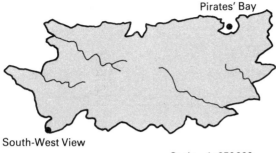

Pirates' Bay

South-West View

Scale 1 : 250 000

3. Erica had two photographs, one an enlargement of the other. The building on the first photograph was 6 cm wide and 4 cm high. On the enlargement the height of the building was 10 cm. How wide was it?

4. A rectangular room is 4.8 m long and 4 m wide. In drawing a plan of the room Mr Cohen drew the width to be 8 cm. What should he draw the length to be?

5. A cylindrical tin has a base radius of 36 mm and a height of 42 mm. Find the height of a similar tin that has a base radius of 96 mm.

21 Equations and Inequations

Exercise 1

Solve these simple equations:

1. $x + 4 = 10$
2. $x - 5 = 9$
3. $x - 9 = 2$
4. $9x = 99$
5. $7x = 84$
6. $\dfrac{x}{5} = 5$
7. $14 - x = 9$

8. $26 - x = 17$
9. $2x + 5 = 21$
10. $3x - 6 = 18$
11. $3x + 6 = 18$
12. $\dfrac{x}{4} = 5$
13. $x \div 7 = 8$
14. $4x - 5 = 23$

15. $3x - 9 = 12$
16. $4x + 3 = 31$
17. $5x - 3 = 32$
18. $2x - 9 = 9$
19. $4x + 4 = 28$
20. $7x - 11 = 52$

Exercise 2

Solve these simple equations:

1. $2p = 15$
2. $5k = 8$
3. $x + 5 = 7\frac{1}{2}$
4. $d - 4 = 8\frac{1}{2}$
5. $v - 2 = 4.5$
6. $3m = 1.5$
7. $\dfrac{l}{2} = 2.5$

8. $t \div 4 = 1\frac{1}{2}$
9. $2a + 1 = 8$
10. $2u - 1 = 10$
11. $2e + 5 = 18$
12. $3h + 2 = 10$
13. $3n + 2 = 6\frac{1}{2}$
14. $5w - 4 = 9$

15. $5f + 1 = 22$
16. $4g - 1 = 9$
17. $3b - 4 = 12$
18. $2q + 2 = 21$
19. $2y - 4 = 13$
20. $6c - 3 = 17$

Exercise 3

Solve these simple equations:

1. $x + 4 = 1$
2. $y + 10 = 6$

3. $m - 1 = {}^-6$
4. $n - 7 = {}^-9$

5. $8 - h = {}^-2$
6. $8 - h = 10$

7. $2d = {}^-8$ **10.** $t \div {}^-2 = 6$ **13.** $3p - 8 = {}^-5$

8. $5e = {}^-15$ **11.** ${}^-2x = 9$ **14.** $4z + 6 = {}^-2$

9. $\dfrac{b}{5} = {}^-1$ **12.** $2v + 7 = 1$ **15.** $2q - 5 = {}^-11$

Exercise 4

For each question, form an equation then solve it:

1. When 4 is added to a number, the answer is 12. Find the number.

2. I think of a number then add 7. If the answer is 16, find the number.

3. I think of a number then subtract 5. If the answer is 11, find the number.

4. If 7 is subtracted from a number, the answer is 9. Find the number.

5. A number, when multiplied by 8, gives 64. Find the number.

6. When a number is multiplied by 11, the answer is 77. Find the number.

7. A number, when divided by 2, gives 3. Find the number.

8. When a number is divided by 8, the answer is 6. Find the number.

9. I have 125 g of sweets. If I buy another x g, I shall have 350 g altogether. Find x.

10. After travelling x miles, Judy must travel a further 87 miles to complete her journey of 213 miles. How many miles has she travelled so far?

11. Find angle $t°$.

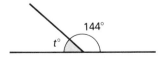

12. Austin saved £v each week for 7 weeks. If he saved £28 altogether, find his weekly savings.

For each question, form an equation then solve it:

1. Fiona thought of a number, doubled it, then added 7. The answer was 15. What number did Fiona think of?

2. Paul thought of a number, multiplied it by 4, then added 5. The result was 17. What number did Paul think of?

3. Seamus thought of a number, multiplied it by 3, then he subtracted 5. The answer was 7. What number did Seamus think of?

4. A number is multiplied by 2 then 7 is subtracted. The result is 9. Find the number.

5. 9 is added to 5 times a certain number. The result is 24. Find the number.

6. If Yvonne were twice her present age she would be 5 years older than Oliver—who is 25 years old. How old is Yvonne?

7. Andrea has 3 times as many sweets as Ben. If she gave 8 away, she would have 13 left. How many sweets does Ben have?

8. I have 3 pieces of wood of equal length. If I had another piece 15 cm long, I should have a total length of 78 cm. How long is each piece?

9. Find d.

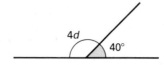

10. Find u.

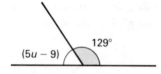

11. Vicki saved a certain amount of money. If she had 6 times as much, she would only need a further £11 to have £65. How much has she saved?

12. Sean travelled by car at a steady v m.p.h. for 2 h. He then needed to travel a further 48 miles to complete a journey of 142 miles. Find the speed, v, in miles per hour.

13. Owen has f 5 p pieces. If he were to have a further 35 p, he would have £1. How many 5 p pieces does he have?

14. Seven people sold d tickets each for a dance. 45 people who bought tickets did not go to the dance. If 200 people attended the dance, how many tickets were sold by each person?

15. A rectangle is l cm long and 8 cm wide. Its perimeter is 30 cm. Find its length.

Exercise 6

Solve these equations:

A
1. $8 - x = 5$
2. $17 - x = 13$
3. $15 - h = 4$
4. $8 - 2p = 2$
5. $14 - 3k = 11$

6. $20 - 2d = 6$
7. $30 - 4y = 10$
8. $48 - 5n = 40$
9. $23 - 3b = 17$
10. $41 - 4v = 17$

11. $34 - 5a = 19$
12. $17 - 2f = 12$
13. $27 - 6z = 3$
14. $12 - 3t = 4$
15. $46 - 8c = 10$

B
1. $5 - t = 8$
2. $14 - c = 19$
3. $15 - u = 17$
4. $15 - 2y = 21$
5. $8 - 3a = 11$

6. $14 - 2w = 26$
7. $2 - 4f = 18$
8. $12 - 3d = 24$
9. $9 - 2p = 25$
10. $12 - 3x = 33$

11. $20 - 5g = 25$
12. $1 - 4k = 33$
13. $16 - 6e = 52$
14. $30 - 7n = 44$
15. $4 - 4m = 24$

Exercise 7

For each question, find whether or not the given value of x satisfies the equation:

1. $3x - 7 = 17$ $(x = 8)$
2. $4x - 9 = 5$ $(x = 3\frac{1}{2})$
3. $15 - 2x = 9$ $(x = 3)$
4. $19 - 3x = 5$ $(x = 4)$
5. $x + 6 = 3x$ $(x = 3)$
6. $3x + 7 = x + 9$ $(x = 1)$

7. $2x - 5 = x + 3$ $(x = 2)$

8. $x - 2 = 14 - 3x$ $(x = 4)$

9. $7 - x = 18 - 4x$ $(x = 3)$

10. $4x - 5 = 10 - 2x$ $(x = 2\frac{1}{2})$

11. $2(x - 7) = 32$ $(x = 9)$

12. $3(2x + 1) = 30$ $(x = 7)$

13. $2(4x - 3) = 18$ $(x = 3)$

14. $6(x - 5) = x$ $(x = 6)$

15. $4(x + 2) = 2x + 14$ $(x = 4)$

16. $5 - 2(x - 2) = 1$ $(x = 4)$

17. $34 - 3(x + 5) = 15$ $(x = 2)$

18. $2(3x - 5) = 16 - (x - 2)$ $(x = 4)$

19. $4(2x + 1) = 28 + (2x + 9)$ $(x = 5\frac{1}{2})$

20. $3(2x - 7) = 18 - 3(x - 2)$ $(x = 5)$

21. $2(4x + 1) = 7 - 2(x - 1)$ $(x = 5)$

22. $5(2x + 3) = 30 + 4(x - 3)$ $(x = \frac{1}{2})$

23. $14 - 3(x - 7) = 19 - 2(12 - x)$ $(x = 9)$

24. $3(4 - x) = 2(x - 1) + 4$ $(x = 2)$

Exercise 8

Solve for x:

1. $4x = x + 9$

2. $x + 4 = 2x$

3. $2x - 5 = x + 2$

4. $5x - 7 = 2x + 11$

5. $x - 6 = 12 - x$

6. $2 + x = 17 - 2x$

7. $10 - 3x = 5x + 6$

8. $3 - x = 7 - 4x$

9. $5x - 11 = 13 - x$

10. $8x + 3 = 9 + 2x$

11. $16 - 3x = 7x - 10$

12. $6x + 5 = 37 - 4x$

Exercise 9

Solve for x:

1. $3(x + 5) = 21$

2. $2(3x - 7) = 22$

3. $5(7x - 4) = 15$

4. $4(2x + 1) = 60$

5. $2(2x - 6) = x$

6. $3(4x - 15) = 7x$

7. $2(x + 4) = 4x + 1$

8. $4(3x - 4) = 6x - 7$

9. $5 + 4(x + 3) = 25$

10. $9 - 2(x - 5) = 13$

11. $7 - 3(x + 1) = 4$

12. $2(x + 2) = 12 + (x - 3)$

13. $3(2x - 1) = 27 - (x + 2)$

14. $4(3x - 2) = 2(5x + 4)$

15. $4 + 3(x - 5) = 7 - (x - 12)$

16. $10 - 3(x - 4) = 2(3x - 4)$

Exercise 10

1. Adrian is twice as old as Bert and Cherie is 7 years older than Bert. Take Bert to be x years of age.
   (*a*) Write an expression for Adrian's age.
   (*b*) Write an expression for Cherie's age.
   (*c*) Write an equation in x, showing that the sum of all three ages is 55 years.
   (*d*) Solve the equation formed in part (*c*), and give the three ages.

2. Tickets for a concert cost £4, £6 and £8 each. Three times more £4 tickets were sold than £6 tickets. If 15 more £8 tickets had been sold, the number of £8 tickets sold would have equalled the number of £6 tickets sold. Let the number of £6 tickets sold be y.
   (*a*) Write an expression for the number of £4 tickets sold.
   (*b*) Write an expression for the number of £8 tickets sold.
   (*c*) Given that the total number of tickets sold was 175, form an equation in y—then solve it. How many tickets of each type were sold?

3. For a concert, 200 tickets were sold. The tickets were £3 and £5 each. £880 was taken from the sale of the tickets.
   (*a*) If z £5 tickets were sold, express in terms of z the number of £3 tickets sold.
   (*b*) Write an expression for the amount of money taken from the sale of the £5 tickets.
   (*c*) Write an expression for the amount of money taken from the sale of the £3 tickets.
   (*d*) Form an equation in z, then solve it to find the number of £5 tickets sold.
   (*e*) How many £3 tickets were sold?

4. The length of a rectangle is 4 cm longer than the breadth.
   (*a*) Write an expression for the length.
   (*b*) If the perimeter is 44 cm, write an equation in terms of b—then solve it to find the length and breadth of the rectangle.

b cm

Linear Inequalities

Exercise 11

A Answer these:

1. Is 10 less than 19?
2. Is 15 greater than 7?
3. Is 29 greater than 31?
4. Is 72 less than 100?
5. Is 568 less than 499?
6. Is 812 greater than 700?

B Answer these:

1. Is ⁻12 less than ⁻8?
2. Is ⁻18 less than 7?
3. Is 4 greater than ⁻2?
4. Is ⁻15 greater than 12?
5. Is 93 less than ⁻87?
6. Is ⁻36 greater than ⁻19?
7. Is ⁻45 greater than ⁻53?
8. Is ⁻100 less than ⁻41?

C Copy each sentence and write 'is greater than' or 'is less than' in place of each question mark:

1. 7 $\boxed{?}$ 9
2. 12 $\boxed{?}$ 21
3. 39 $\boxed{?}$ 31
4. 698 $\boxed{?}$ 600
5. 780 $\boxed{?}$ 899
6. 1001 $\boxed{?}$ 1010

D Copy each sentence and write 'is greater than' or 'is less than' in place of each question mark:

1. ⁻9 $\boxed{?}$ ⁻2
2. ⁻9 $\boxed{?}$ 2
3. 12 $\boxed{?}$ ⁻18
4. ⁻41 $\boxed{?}$ ⁻39
5. ⁻54 $\boxed{?}$ ⁻60
6. ⁻100 $\boxed{?}$ ⁻80
7. ⁻467 $\boxed{?}$ ⁻510
8. ⁻2120 $\boxed{?}$ ⁻2102

Exercise 12

Write whether each statement is true or false:

A
1. $15 > 19$
2. $28 < 46$
3. $39 > 17$
4. $17 < 39$
5. $42 > 30$
6. $30 < 42$
7. $56 < 35$
8. $35 > 56$
9. $184 > 190$
10. $272 < 290$
11. $307 > 310$
12. $524 < 542$

B 1. $^-15 < ^-19$ **5** $37 < ^-49$ **9.** $^-205 < ^-210$

2. $^-19 < ^-15$ **6.** $^-49 < 37$ **10.** $^-510 < ^-508$

3. $^-18 < 12$ **7.** $^-51 > ^-39$ **11.** $^-702 > ^-700$

4. $12 > ^-18$ **8.** $^-84 > ^-90$ **12.** $^-964 > ^-976$

C 1. (a) $13 < 17$ (b) $^-13 < ^-17$ (c) $^-17 > ^-13$

2. (a) $15 > 12$ (b) $^-15 > ^-12$ (c) $^-15 < ^-12$

3. (a) $21 > 17$ (b) $^-21 > ^-17$ (c) $^-21 < ^-17$

4. (a) $^-45 < ^-30$ (b) $45 > 30$ (c) $^-30 > ^-45$

5. (a) $^-36 > 19$ (b) $^-19 > 36$ (c) $^-19 < ^-36$

Exercise 13

Copy these, but replace each question mark with $<$ or $>$ to make each statement true:

A 1. $6 \boxed{?} 12$ **5.** $84 \boxed{?} 90$ **9.** $766 \boxed{?} 749$

2. $4 \boxed{?} 19$ **6.** $91 \boxed{?} 100$ **10.** $2020 \boxed{?} 2002$

3. $23 \boxed{?} 17$ **7.** $110 \boxed{?} 101$ **11.** $3190 \boxed{?} 3910$

4. $47 \boxed{?} 33$ **8.** $641 \boxed{?} 614$ **12.** $4080 \boxed{?} 4800$

B 1. $^-9 \boxed{?} ^-7$ **5.** $34 \boxed{?} ^-19$ **9.** $408 \boxed{?} ^-98$

2. $^-6 \boxed{?} ^-8$ **6.** $^-50 \boxed{?} 48$ **10.** $712 \boxed{?} ^-835$

3. $^-12 \boxed{?} ^-15$ **7.** $^-86 \boxed{?} 88$ **11.** $^-69 \boxed{?} ^-152$

4. $^-20 \boxed{?} 10$ **8.** $^-140 \boxed{?} ^-135$ **12.** $^-409 \boxed{?} ^-410$

Exercise 14

Write each of the following in another way.
(Whenever $<$ is used, use $>$. Whenever $>$ is used, use $<$.)

e.g. $7 < 12$ so $12 > 7$.

1. $14 > 9$ 4. $79 > 50$ 7. $^-8 < ^-3$ 10. $^-30 < ^-20$

2. $19 < 22$ 5. $25 < 60$ 8. $^-4 > ^-11$ 11. $^-51 < 0$

3. $44 > 32$ 6. $108 < 110$ 9. $^-16 > ^-21$ 12. $0 > ^-26$

A The natural numbers that are less than 9 are shown on the *number line*:

1. What numbers on the line are less than 5?

2. What numbers on the line are greater than 6?

3. What numbers on the line are less than or equal to 6?

4. Which number on the line is 3 greater than 2?

5. Which number on the line is 2 less than 7?

B For each question, make a separate copy of the number line given and mark your answers on the line. Use dots (as in part A) to show your answers.

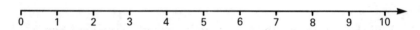

Show the set of numbers that are:

1. less than 5	**6.** smaller than 3
2. less than or equal to 7	**7.** more than 4
3. greater than 9	**8.** between 3 and 6
4. greater than 2	**9.** between 7 and 9
5. greater than or equal to 6	**10.** between 2 and 8

C For each question, make a separate copy of the number line given and mark your answers on the line:

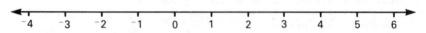

Show the set of integers that are:

1. less than ⁻2	**6.** less than or equal to ⁻1
2. more than ⁻1	**7.** less than or equal to 3
3. less than 4	**8.** greater than or equal to 2
4. greater than 0	**9.** between ⁻3 and 5
5. less than 0	**10.** between ⁻2 and ⁺2

280

Exercise 16

For each question, write all possible answers:

1. I have less than 10p in my pocket. How much might it be?

2. The only coins I have are £1 coins. If I have less than £6, how much might I have?

3. There are less than 8 pieces of chalk in a box. How many pieces might there be in the box?

4. It can take up to 6 min to answer a certain question. How long, to the nearest minute, could it have taken to answer the question?

5. To drive a car, you must be 17 years of age or older, how old can Erika be (to the nearest year) if she is not allowed to drive?

Linear Inequations

Exercise 17

1. $\boxed{?} < 5$
   Will the statement above be true if the box is replaced by:
   (a) 4?   (b) 2?   (c) 0?   (d) 5?   (e) 9?   (f) ⁻3?

2. $\boxed{?} \geqslant 4$
   Will the statement above be true if the box is replaced by:
   (a) 7?   (b) 3?   (c) 0?   (d) 4?   (e) 10?   (f) ⁻2?

3. $x \leqslant 6$
   Will the statement above be true if x equals:
   (a)4?   (b) 6?   (c) 0?   (d) 7?   (e) ⁻3?   (f) 1?

4. $y > 2$
   Will the statement above be true if y equals:
   (a) 1?   (b) 6?   (c) 0?   (d) ⁻3?   (e) 8?   (f) 5?

5. $d \geqslant {}^-3$
   Will the statement above be true if d equals:
   (a) 1?   (b) 0?   (c) ⁻2?   (d) ⁻3?   (e) ⁻5?   (f) 8?

6. $p < {}^-5$
   Will the statement above be true if p equals:
   (a) 0?   (b) 4?   (c) ⁻5?   (d) ⁻7?   (e) 3?   (f) ⁻10?

Consider the *inequation* $x < 6$. If x is a natural number, possible values of x are 1, 2, 3, 4, 5.

$\{1, 2, 3, 4, 5\}$ is called the *solution set* of $x < 6$.

Exercise 18

Throughout this exercise, take x to be a natural number. Find the solution set for each of the given inequations. When instructed, show the solution set on a number line.

e.g. 1 $x < 4$

     Solution set $= \{1, 2, 3\}$

e.g. 2 $x \geqslant 6$

     Solution set $= \{6, 7, 8, 9, \ldots\}$

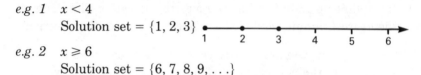

Note   The arrows suggest that the number lines continue. In e.g.2, 9 is marked with a dot to show that it is in the solution set. The arrow next to 9 means that all numbers after 9 are marked in the same way as 9, with a dot, and therefore are in the solution set.

1. $x < 5$ **4.** $x < 7$ (Show answers on a number line.)

2. $x \leqslant 4$ **5.** $x > 8$ (Show answers on a number line.)

3. $x > 2$ **6.** $x \leqslant 10$ (Show answers on a number line.)

Exercise 19

A Throughout this exercise, take x to be a natural number. Find the solution set for each of the given inequations.

e.g. 1 $x + 3 \leqslant 8$ Compare the equation $x + 3 = 8$

         $x \leqslant 5$ $\underline{\underline{x = 5}}$

     Solution set $= \{1, 2, 3, 4, 5\}$

e.g. 2 $x - 5 > 7$ Compare the equation $x - 5 = 7$

         $x > 12$ $\underline{\underline{x = 12}}$

     Solution set $= \{13, 14, 15, \ldots\}$

1. $x + 4 < 7$

2. $x - 5 < 4$

3. $x - 1 > 8$

4. $x + 6 > 7$

5. $x + 9 \geqslant 16$

6. $x + 3 \leqslant 11$

7. $x - 7 \leqslant 0$

8. $x - 6 < 0$

9. $x + 8 > 12$

10. $x - 3 < 9$

B Show, on a number line, the answers to part A, questions 1, 6 and 9.

Exercise 20

A Where x is a natural number, find the solution set for each of these inequations:

e.g. 1 $3x > 15$
$ x > 5$

e.g. 2 $4x < 17$
$ x < 4.25$

     Solution set $= \{6, 7, 8, 9, \ldots\}$ Solution set $= \{1, 2, 3, 4\}$

1. $2x < 16$

2. $2x \leqslant 13$

3. $3x \geqslant 12$

4. $4x > 24$

5. $3x \leqslant 27$

6. $5x \geqslant 13$

7. $6x < 15$

8. $8x > 56$

9. $2x > 23$

10. $5x < 55$

11. $3x > 32$

12. $4x \leqslant 27$

B Show, on a number line, the answers to part A, questions 2, 3, 7 and 9.

Exercise 21

1. $x + 5 < 9$:

     (*a*) Give two possible whole number values of x.

     (*b*) Give a value of x that is not an integer.

2. $x - 5 \geqslant 7$:

     (*a*) Give one possible whole number value of x.

     (*b*) Give a value of x that is not a natural number.

3. $x + 8 \leqslant 10$:

     (*a*) Give a whole number value of x.

     (*b*) Give a value of x that is not an integer.

4. Give a rational number value of x so that $x + 3 < 4$.

5. Give a rational number value of x so that $2x \leqslant 5$.

6. If $3x > 15$ give one possible value of x.

7. Give a value of x so that $4x < 4$.

8. Give a value of x so that $4x \leqslant 3$.

9. If $2x < 10$, give one irrational number that satisfies the inequation.

10. Give a value of x so that $5x \geqslant 9$.

Exercise 22

1. $2x \leqslant 8$ and $x > 2$:
(a) Give two possible values of x that are natural numbers.
(b) Give a value of x that is not an integer.

2. Give one possible value of x so that $3x > 6$ and $x < 5$.

3. Give one possible value of x so that $3x > 6$ and $x < 3$.

4. Give one possible value of x so that $4x \leqslant 9$ and $x > 1$.

5. Give a non-integral* value of x so that $2x < 10$ and $x \geqslant 4$.

6. Write a value of x so that $x < 9$ and $4x > 32$.

7. Give a value of x so that $x + 4 < 8$ and $x > 3$.

8. Give two values of x so that $5x \leqslant 25$ and $x - 3 > 1$.

9. Give a value of x so that $x + 3 > 3$ and $x - 4 < 3$.

10. Give three values of x so that $x + 8 < 15$ and $3x \geqslant 15$.

*See the glossary, p.479.

284

22 Co-ordinates and Graphs

Exercise 1   Co-ordinates

1. In the diagram, A is the point (8, 0).

   (a) What are the co-ordinates of B?

   (b) Q is the mid-point of OB. What are the co-ordinates of Q?

   (c) P is the mid-point of OA. What are the values of x and y at P?

   (d) What are the co-ordinates of the mid-point of AB?

   (e) What are the co-ordinates of the mid-point of PQ?

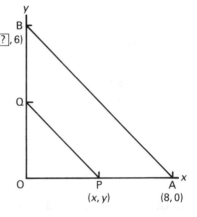

2. In the diagram, A is the point (8, 6) and B the point (4, 4).

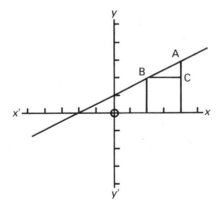

   (a) What are the co-ordinates of C?

   (b) At which point does the line AB produced cut the x-axis?

   (c) At which point does the line AB produced cut the y-axis?

3. Draw a pair of axes as shown and draw the graph of $x = 6$ using a broken line.

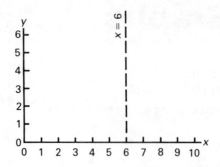

(a) Plot the points A(6, 6), B(3, 6), C(6, 4), D(2, 4), E(2, 2) and F(6, 2).

(b) Join A to B to C to D to E to F using straight lines.

(c) Complete the figure so that it is symmetrical about AF.

(d) Calculate the area of the completed figure.

Straight-line Graphs

Exercise 2

Draw a pair of axes as shown.
(Use a scale of 2 cm to 1 unit.)

Draw the following graphs using the one pair of axes.

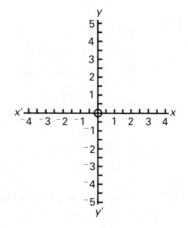

1. $x = 3$

2. $x = {}^-1$

3. $x = {}^-4$

4. $x = 2.5$

5. $x = 1.7$

6. $y = 1$

7. $y = {}^-2$

8. $y = 3.5$

9. $y = 0.8$

10. $y = {}^-1.2$

286

Exercise 3

1. Draw a pair of axes as shown. x should range from $^-2$ to 4 ($^-2 \leqslant x \leqslant 4$). Use a scale of 2 cm to 1 unit.
   y should range from $^-10$ to 10 ($^-10 \leqslant y \leqslant 10$). Use a scale of 1 cm to 1 unit.

2. Plot the graph of $y = 3x - 5$.

3. Using the same pair of axes, plot the graph of $y = x + 1$.

4. Find the co-ordinates of the point of intersection of the two graphs.

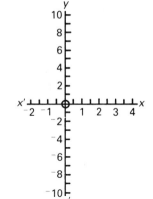

Exercise 4

1. Draw a pair of axes as shown. x and y should both range from 0 to 8. Use a scale of 1 cm to 1 unit on each axis.

2. Consider the equation

   $$x + y = 5$$

   (a) When $x = 3$, $y = 2$. Plot the point $(3, 2)$.
   (b) When $x = 0$, $y = 5$. Plot the point $(0, 5)$.
   (c) Work out another point where $x + y = 5$, then plot it.
   (d) Draw the graph of $x + y = 5$.

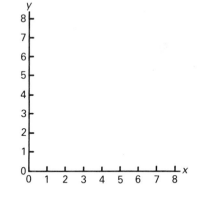

3. (a) Using the same pair of axes, draw the graph of $y = 5 - x$.
   (b) Write what you notice.

4. Using the same pair of axes, draw these graphs:
   (a) $x + y = 3$ (b) $x + y = 8$

287

Exercise 5

Draw a pair of axes where the x-values range from $^-4$ to 12 and the y-values from $^-2$ to 7. (Use a scale of 1 cm to 1 unit on both axes.)

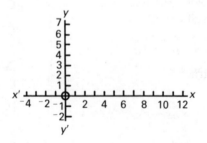

Consider the equation $x + 2y = 10$.

1. (a) When $x = 0$, find y.
    (b) Plot the point $(0, y)$, where y is the value calculated in part (a).
2. (a) When $y = 0$, find x.
    (b) Plot the point $(x, 0)$, where x is the value calculated in part (a).

3. (a) When $x = 4$, find y.   (b) Plot $(4, y)$.

4. (a) When $x = 5$, find y.   (b) Plot $(5, y)$.

5. (a) When $x = ^-4$, find y.   (b) Plot $(^-4, y)$.

6. Join the plotted points using a straight line and label the line with its equation, $x + 2y = 10$.

You have probably realised that it is not necessary to plot as many points as in Exercise 4 to be able to draw a straight line graph. Although only 2 points are needed, it is better to plot 3 points (the third point serves as a check).

Note   Usually, the easiest points to calculate and plot are when $x = 0$, and $y = 0$.

For example, if $2x + 3y = 12$
When $y = 0$, $2x$ $= 12$
          so x $= 6$;   giving the point $(6, 0)$.
When $x = 0$, $3y = 12$
          so $y = 4$;   giving the point $(0, 4)$.

(Both these points lie on the axes and are easy to find.)

288

Exercise 6

A For each question, draw a pair of axes as shown, where the x-values range from 0 to 9 and the y-values range from 0 to 12. (Use a scale of 1 cm to 1 unit.)

Draw the following graphs:

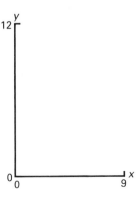

1. $2x + y = 10$ **5.** $x + 3y = 9$
2. $2x + y = 12$ **6.** $2x + 3y = 18$
3. $4x + y = 12$ **7.** $3x + 4y = 12$
4. $x + 2y = 8$ **8.** $5x + 3y = 30$

B For each question, draw a pair of axes as shown, where the x-values range from 0 to 9 and the y-values range from $^-6$ to 9. (Use a scale of 1 cm to 1 unit on both axes.)

Draw the following graphs:

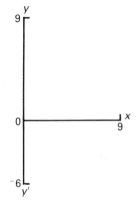

1. $x - y = 3$
2. $2x - y = 5$
3. $x - 3y = 6$

C For each question, draw a pair of axes as shown, where the x-values range from $^-6$ to 2 and the y-values range from $^-7$ to 7. (Use a scale of 1 cm to 1 unit on both axes.)

Draw the following graphs:

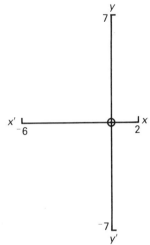

1. $y = x + 5$
2. $x - y = ^-4$
3. $x + y = ^-5$

D Draw a pair of axes, where the *x*-values range from ⁻3 to 5 (use a scale of 1 cm to 1 unit) and the *y*-values range from ⁻15 to 10 (use a scale of 1 cm to 2 units on the *y*-axis).

1. Draw the graph of $3x - y = 5$ and label it.

2. Using the same pair of axes, draw the graph of $3x + y = 1$. Label this graph.

3. Find the co-ordinates of the point of intersection of the two graphs.

Exercise 7

1. The diagram shows a sketch of the graph of $y = 2x - 7$. Calculate the co-ordinates of:
(*a*) A   (*b*) B

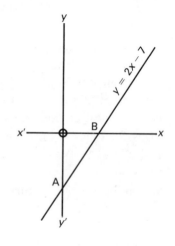

2. The diagram shows a sketch of the graph of $x + 3y = 6$. Calculate the co-ordinates of:
(*a*) P   (*b*) Q

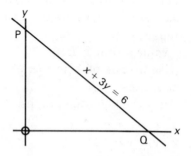

3. The graph of $4x + 3y = 24$ cuts the *x*-axis at L and the *y*-axis at M. Calculate the co-ordinates of:
(*a*) L   (*b*) M

4. Mrs Charnley bought x tickets for a concert costing £3 each and y tickets costing £5 each:

(a) Write an expression for the total cost of the £3 tickets.

(b) Write an expression for the total cost of the £5 tickets.

(c) Write an expression for the total cost of all the tickets.

(d) Write an equation involving both x and y, if the total cost of the tickets was £45.

(e) Using the equation found in part (d), find y when $x = 0$.

(f) Using the equation found in part (d), find x when $y = 0$.

(g) Draw a pair of axes as shown. (Use a scale of 1 cm to 1 unit.) Draw the graph of the equation found in part (d).

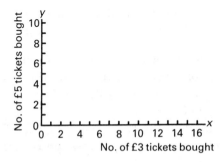

(h) From the graph, find a set of ordered pairs that satisfies the problem.

Exercise 8

1. (a) Draw a pair of axes where the x-values range from $^-3$ to 5 and the y-values range from $^-2$ to 10. (Use a scale of 1 cm to 1 unit on both axes.)

(b) Plot the graph of $y = (3x + 5)/2$. (Use the table on p.207 to help you.)

(c) At which point does the graph cross the y-axis?

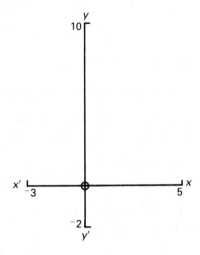

2. (a) Draw a pair of axes where the x-values range from ⁻8 to 7 (use a scale of 1 cm to 2 units) and the y-values range from ⁻2 to 4 (use a scale of 2 cm to 1 unit).
   (b) Plot the graph of $y = 2(x + 3)/5$. (Use the table on p.208 to help you.)
   (c) At which point does the graph cross the x-axis?
   (d) At which point does the graph cross the y-axis?

3. (a) Draw a pair of axes where the x-values range from ⁻4 to 8 (use a scale of 1 cm to 1 unit) and the y-values range from ⁻10 to 10 (use a scale of 1 cm to 2 units).
   (b) Plot the graph of $y = 3(x - 2)/2$.
   (c) At which point does the graph cross the y-axis?
   (d) At which point does the graph cross the x-axis?

4. (a) Draw a pair of axes where the x-values range from ⁻4 to 8 (use a scale of 1 cm to 1 unit) and the y-values range from ⁻10 to 15 (use a scale of 2 cm to 5 units).
   (b) Plot the graph of $y = (4x - 3)/2$.
   (c) At which point does the graph cross the y-axis?
   (d) At which point does the graph cross the x-axis?

Exercise 9

1. Does $(2, 8)$ lie on the graph of $y = 3x + 2$?

2. Does $(3, ⁻2)$ lie on the graph of $y = 2x - 6$?

3. Does $(0, 2)$ lie on the graph of $y = 5x - 2$?

4. Does $(⁻2, 3)$ lie on the graph of $y = 2x + 7$?

5. Does $(6, 14)$ lie on the graph of $y = 4x - 10$?

6. The straight line $y = 2x + 5$ passes through the point $(1, y)$. Find the value of y.

7. The straight line $y = 3x - 8$ passes through the point $(x, 4)$. Find the value of x.

8. The straight line $y = 3x + c$ passes through the point $(2, 9)$. Find the value of c.

9. At which point does the line $y = 2x - 7$ cut the y-axis?

10. At which point does the line $y = 4x - 12$ cut the x-axis?

11. The straight line $y = 2x + c$ passes through the point $(3, 8)$. Find the value of c.

12. The straight line $y = mx - 5$ passes through the point $(4, 7)$. Find the value of m.

13. The graph of $y = 4x - 1$ passes through the point $(2, y)$. Find the value of y.

14. The graph of $y = 5x + c$ passes through the point $(^-2, {}^-3)$. Find the value of c.

15. The graph of $y = 2x + 9$ passes through the point $(x, 5)$. Find the value of x.

Exercise 10   Graphical Solutions of Simple Linear Equations

A 1. Solve the equation $2x - 6 = 0$.

2. (*a*) Draw a pair of axes as shown.
       x should range from $^-2$ to 6 and y from $^-12$ to 8.
   (*b*) Draw the graph of $y = 2x - 6$.

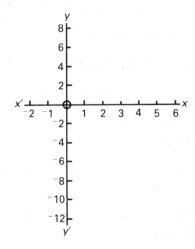

3. Consider the equation $2x - 6 = 0$. If $y = 2x - 6$, then $y = 0$. Now $y = 0$ is the x-axis and $y = 2x - 6$ is the graph that has been drawn.
   (a) At which point does the graph of $y = 2x - 6$ cross the x-axis?
   (b) Compare the x-co-ordinate of the point in part (a) with the solution to the equation in question 1 and write what you notice.

B 1. Solve the equation $2x = 6$.

2. Consider the equation $2x - 6 = 0$:

$$2x = 6 \quad \text{has the same solution.}$$

Now if $y = 2x$ then $y = 6$.
So using the same pair of axes draw the graphs of $y = 2x$ and $y = 6$.

3. Write the x-value at the point of intersection of the two graphs.

4. Compare the answers to questions 1 and 3, then write what you notice.

C Solve graphically:
   1. $2x - 4 = 0$
   2. $x + 3 = 0$
   3. $2x - 3 = 6$
   4. $3x + 5 = 11$

Gradient

The *gradient* of a line is a measure of how steep it is.

The steepness depends on how much the line rises vertically (the rise) compared with either its length or the horizontal distance (the tread).

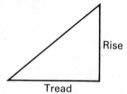

In Mathematics we use:

$$\text{Gradient} = \frac{\text{rise}}{\text{tread}}$$

The gradient of any straight line can be found by drawing a right-angled triangle under the line then dividing the rise by the tread. This is shown in the following examples.

e.g. 1 Find the gradient of the straight line joining the points (2, 7) and (5, 13).

Method 1 From the diagram,

$$\text{the rise} = 13 - 7 = 6$$

$$\text{the tread} = 5 - 2 = 3$$

$$\text{Gradient} = \frac{\text{rise}}{\text{tread}} = \frac{6}{3} = \underline{\underline{2}}$$

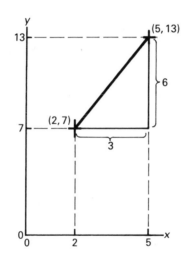

The gradient of a line joining two points whose co-ordinates are known can also be found from a formula:

Gradient of a line $= \dfrac{(y_2 - y_1)}{(x_2 - x_1)}$

where the two points are (x_1, y_1) and (x_2, y_2).

Method 2 Let $(x_1, y_1) = (2, 7)$ and $(x_2, y_2) = (5, 13)$.

$$\text{Gradient} = \frac{(y_2 - y_1)}{(x_2 - x_1)} = \frac{(13 - 7)}{(5 - 2)} = \frac{6}{3} = \underline{\underline{2}}$$

e.g. 2 Find the gradient of the straight line joining the points $(^-3, 14)$ and $(1, 2)$.

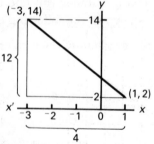

Method 1   From the diagram,

$$\text{the rise} = {}^-12$$

$$\text{the tread} = 4$$

$$\text{Gradient} = \frac{\text{rise}}{\text{tread}} = \frac{{}^-12}{4} = \underline{\underline{{}^-3}}$$

(*Note*   The gradient is negative because the line slopes downwards.)

Method 2   Let $(x_1, y_1) = (^-3, 14)$ and $(x_2, y_2) = (1, 2)$.

$$\text{Gradient} = \frac{(y_2 - y_1)}{(x_2 - x_1)} = \frac{(2 - 14)}{(1 - {}^-3)} = \frac{{}^-12}{4} = \underline{\underline{{}^-3}}$$

Exercise 11

A Find the gradient of each of the line segments given:

1.

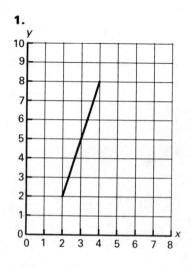

2.

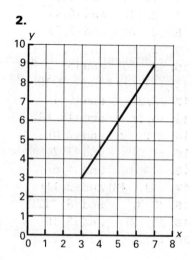

3.   **4.**

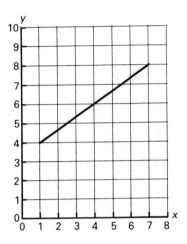 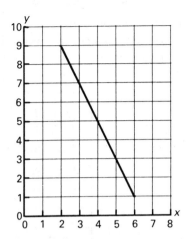

B Find the gradient of the straight line segments joining each given pair of points.

1. (5, 9)   and (7, 17)
2. (4, 8)   and (8, 20)
3. (1, 6)   and (4, 21)
4. (8, 9)   and (10, 13)
5. (2, 6)   and (4, 20)
6. (15, 21)   and (21, 27)
7. (2, 10)   and (5, 4)
8. (3, ⁻5)   and (5, 7)
9. (2, ⁻6)   and (5, ⁻15)
10. (2, ⁻11)   and (5, ⁻2)
11. (⁻1, 5)   and (⁻3, 19)
12. (3, 2)   and (7, 4)
13. (5, 1)   and (13, 13)
14. (⁻2, ⁻2)   and (2, 0)
15. (1, 9)   and (7, 24)
16. (⁻4, 10)   and (4, 6)
17. (⁻1, ⁻13)   and (2, 11)
18. (3, 3)   and (8, ⁻12)
19. (0, 12)   and (3, 15)
20. (⁻3, 7)   and (2, 2)

297

A **1.** Draw a pair of axes as shown. Use a scale of 2 cm to 1 unit on the x-axis and 1 cm to 1 unit on the y-axis. x ranges from 0 to 6 and y from 0 to 20.

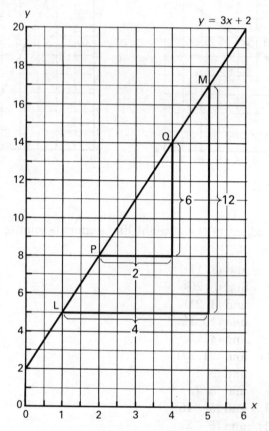

2. Draw the graph of $y = 3x + 2$.

3. Mark two points on your graph. (Choose two points at which it is easy to find the co-ordinates, such as points P and Q in the diagram.) Find the gradient of the line by drawing a right-angled triangle under the graph (as shown).

4. Repeat question 3 for two other points such as L and M in the diagram.

5. Repeat question 3 for a third pair of points.

6. Write what you notice about the gradients calculated from the three different right-angled triangles.

7. Compare the gradients found in questions 3, 4 and 5 with the equation of the line ($y = 3x + 2$ in this case). Write what you notice.

B **1.** Draw a pair of axes where x ranges from $^-4$ to 4 (use a scale of 2 cm to 1 unit) and y ranges from $^-5$ to 15 (use a scale of 1 cm to 1 unit).

2. Draw the graph of $y = 2x + 4$.

3–7. Repeat part A (questions 3 to 7) for the graph of $y = 2x + 4$.

Exercise 13

The equations of several graphs are given. For each one, write its gradient.

1. $y = 2x$

2. $y = 3x - 2$

3. $y = 3x + 5$

4. $y = 6x + 1$

5. $y = 7x - 4$

6. $y = 4x + 6$

7. $y = 8x - 9$

8. $y = 5x - 5$

9. $y = \frac{1}{2}x + 8$

10. $y = \frac{3}{4}x - 2$

11. $y = \frac{5x}{2} - 8$

12. $y = \frac{5x}{3} + 4$

Exercise 14

1. (a) Copy and complete the table for the function $y = 5 - 2x$:

	x	$^-4$	$^-2$	0	2	4	
+	^-2x	8			$^-4$		+ ⎫ These two rows are added to give the
+	5		5		5		+ ⎬ values of y in the last row.
	$y = 5 - 2x$	13			1		⎭

(b) Draw a pair of axes where x ranges from $^-4$ to 4 and y ranges from $^-5$ to 15.

2. Draw the graph of $y = 5 - 2x$.

3–7. Repeat Exercise 12, part A, questions 3 to 7, for the graph of $y = 5 - 2x$.

Exercise 15

The equations of several graphs are given. For each one, write its gradient.

1. $y = 6 - 4x$ **5.** $y = {}^-3x + 2$ **9.** $y = {}^-3 + 4x$

2. $y = 7 - 3x$ **6.** $y = {}^-5x - 6$ **10.** $y = 2 - \frac{1}{2}x$

3. $y = 4 - x$ **7.** $y = {}^-4x + 2$ **11.** $y = \dfrac{{}^-3x}{2} + 6$

4. $y = 6 + 3x$ **8.** $y = 2 - 4x$ **12.** $y = 7 - \dfrac{2x}{3}$

Exercise 16

Throughout this exercise, use the scale of 1 cm to 1 unit on each axis.

A 1. Draw a pair of axes where x ranges from $^-3$ to 3 and y ranges from $^-9$ to 9.

2. Plot the graph of $y = 3x$.

3. On the same pair of axes, draw the graph of $2y = 6x$.

4. Write what you notice and explain your answer.

B 1. Draw a pair of axes where x ranges from $^-3$ to 5 and y ranges from $^-9$ to 7.

2. Plot the graph of $y = 2x - 3$.

3. On the same pair of axes, draw the graph of $3y = 6x - 9$.

4. Write what you notice and explain your answer.

C 1. Draw a pair of axes where x ranges from $^-2$ to 4 and y ranges from $^-10$ to 8.

2. Plot the graph of $y = 2 - 3x$.

3. On the same pair of axes, draw the graph of $3x + y = 2$.

4. Write what you notice and explain your answer.

To find the gradient of a graph, always write the equation in the form $y =$ 'something in terms of x'. In that form, the coefficient of x is the gradient.

> The gradient of a straight-line graph of the form $y = mx + c$ is m, the coefficient of x.

Exercise 17

The equations of several graphs are given. For each one rewrite in the form $y = mx + c$, then give its gradient.

1. $2y = 8x$

2. $3y = 6x + 12$

3. $5y = 15x - 5$

4. $4y = 2x + 8$

5. $2y = {}^-6x$

6. $4y = {}^-8x + 20$

7. $3y = 18 - 6x$

8. $3y = 21 - 3x$

9. $5y = 20 - x$

10. $x + y = 4$

11. $2x + y = 7$

12. $4x + y = 12$

13. $12x + 3y = 42$

14. $6x + 2y = 9$

15. $y - 6x = 8$

Quadratic Graphs

Exercise 18

Draw a pair of axes as shown, where
${}^-4 \leqslant x \leqslant 4$ and $0 \leqslant y \leqslant 100$.
Use a scale of 2 cm to 1 unit on the x-axis and 1 cm to 5 units on the y-axis.
Using the one pair of axes, plot the following graphs:

1. $y = x^2$

2. $y = 2x^2$

3. $y = 5x^2$

4. $y = 6x^2$

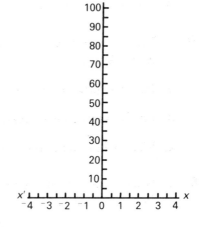

Exercise 19

Draw a pair of axes as shown
where $^-4 \leqslant x \leqslant 4$
and $^-90 \leqslant y \leqslant 10$.
Use a scale of 2 cm to 1 unit on the
x-axis and 1 cm to 5 units on the
y-axis.
Plot all the graphs in this exercise
using the one pair of axes.

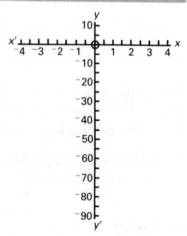

1. (a) Copy and complete the following table for the function
$y = {}^-x^2$.

x	$^-4$	$^-3$	$^-2$	$^-1$	$^-0.5$	0	0.5	1	2	3	4
x^2		9					0.25		4		
$y = {}^-x^2$	$^-16$						$^-0.25$		$^-4$		$^-16$

(b) Plot the graph of $y = {}^-x^2$.

2. Plot the graphs:
(a) $y = {}^-2x^2$ (b) $y = {}^-4x^2$ (c) $y = {}^-5x^2$

Exercise 20

Draw a pair of axes as shown
where $^-8 \leqslant x \leqslant 8$
and $^-25 \leqslant y \leqslant 75$.
Use a scale of 1 cm to 1 unit on the
x-axis and 1 cm to 5 units on the
y-axis.
Plot all the graphs in this exercise
using the one pair of axes.

1. Plot the graph of $y = x^2$.

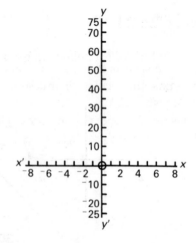

302

2. Copy and complete the following table for the graph of $y = x^2 + 9$.

x	$^-8$	$^-7$	$^-6$	$^-5$	$^-4$	$^-3$	$^-2$	$^-1$	0	1	2	3	4	5	6	7	8
x^2	64					9	4								36		
$y = x^2 + 9$	73			34		18		10								58	

(b) Plot the graph of $y = x^2 + 9$.

3. (a) Copy and complete the following table for the graph of $y = x^2 - 9$.

x	$^-8$	$^-7$	$^-6$	$^-5$	$^-4$	$^-3$	$^-2$	$^-1$	0	1	2	3	4	5	6	7	8
x^2	64				16					1			16				
$y = x^2 - 9$	55				7					$^-8$	$^-5$					40	

(b) Plot the graph of $y = x^2 - 9$.
(c) At which points does the graph cross the x-axis?

4. (a) Plot the graph of $y = x^2 - 25$.
(b) At which points does the graph cross the x-axis?

5. Plot the graphs of:
(a) $y = x^2 + 5$ (b) $y = x^2 - 15$ (c) $y = x^2 - 3$

Linear and Quadratic Graphs

Exercise 21

1. Which of the following graphs could be the graph of $y = 3$?

A   B   C

D **E** **F**

2. The equations of the following nine graphs are:

$$y = 4 \qquad y = x + 4 \qquad y = x^2$$
$$x = 4 \qquad y = x - 4 \qquad y = x^2 - 4$$
$$y = 4x \qquad x + y = 4 \qquad y = -x^2$$

Which is the equation of:

(a) graph A? (d) graph D? (g) graph G?
(b) graph B? (e) graph E? (h) graph H?
(c) graph C? (f) graph F? (i) graph I?

A **C** **E**

B **D** **F**

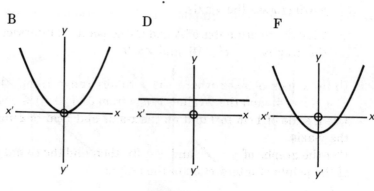

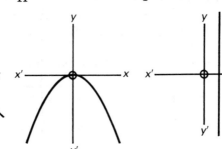

G   H   I

Exercise 22

1. The diagram shows the graphs of $y = x^2 - 16$ and $y = 9$.

   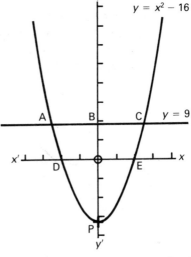

   (a) Draw the graphs. Use a scale of 1 cm to 2 units on the x-axis and 1 cm to 5 units on the y-axis.

   (b) Give the co-ordinates of D and E, the points where $y = x^2 - 16$ crosses the x-axis.

   (c) Give the co-ordinates of the *minimum point* of the graph of $y = x^2 - 16$.

   (d) Give the co-ordinates of B, the point at which the graph of $y = 9$ crosses the y-axis.

   (e) Give the co-ordinates of A and C, the points of intersection of the graphs $y = x^2 - 16$ and $y = 9$.

2. Draw a pair of axes where the x-values range from $^-4$ to 4 ($^-4 \leqslant x \leqslant 4$) and the y-values range from 0 to 20 ($0 \leqslant y \leqslant 20$). Use a scale of 1 cm to 1 unit on the x-axis and 1 cm to 2 units on the y-axis.
   Plot the graphs of $y = x^2$ and $y = 3x$ then find the co-ordinates of the points of intersection of the graphs.

305

3. Draw another pair of axes as for question 2.
   Draw the graphs of $y = x^2$ and $y = x + 6$, then find the co-ordinates of the points of intersection of the graphs.

4. Draw another pair of axes as for question 2.
   Draw the graphs of $y = x^2$ and $2x + y = 8$, then find the co-ordinates of the points of intersection of the graphs.

5. Draw a pair of axes where the x-values range from $^-5$ to 3 ($^-5 \leqslant x \leqslant 3$) and the y-values range from $^-5$ to 25 ($^-5 \leqslant y \leqslant 25$). Use a scale of 1 cm to 1 unit on the x-axis and 2 cm to 5 units on the y-axis.
   Draw the graphs of $y = x^2 - 4$ and $3x + y = 6$, then find the co-ordinates of the points of intersection of the graphs.

Revision Exercises XVI to XXII

Revision Exercise XVI

1. If $y = 3x + 5$, find the value of y when:

(a) $x = 4$ (b) $x = 0$ (c) $x = 7$ (d) $x = {}^-1$ (e) $x = {}^-4$

2. If $y = 7x^2$, find the value of y when:

(a) $x = 0$ (b) $x = 5$ (c) $x = {}^-2$ (d) $x = {}^-1$ (e) $x = \frac{1}{2}$

3. Copy and complete the diagram for the mapping

$x \longrightarrow 3x - 7$:

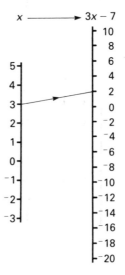

4. Copy and complete the following ordered pairs for the function $y = x^2 + 9$:

$(0, \boxed{?})$, $(2, \boxed{?})$, $({}^-3, \boxed{?})$, $(3, \boxed{?})$

5. Copy and complete the following table for the function $y = x^2 - 10$:

x	$^-3$	$^-2$	$^-1$	0	1	2	3	4	5
x^2						4			
$y = x^2 - 10$			$^-9$						

307

6. Copy and complete either mapping diagram A or B for the function $y = 3(x - 4)/2$:

A

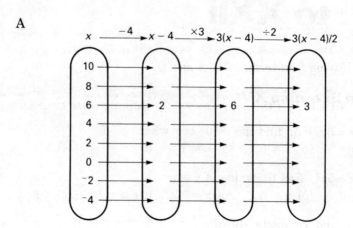

B

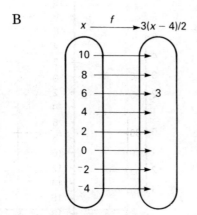

Revision Exercise XVII

1. $V = Ah$ gives the volume of a cylinder of perpendicular height h units where the area of its base is A square units.

Find the volume of a cylinder having a base area of $48\,\text{cm}^2$ and a perpendicular height of $8\,\text{cm}$.

2. (a) If 4 identical books cost £20, find the cost of one book.
(b) If n identical books cost £20, find the cost of one book.
(c) If n identical books cost £c, find the cost of one book.

3. The cost of running a boat-hire firm is £C per day. The firm makes n hirings in a day at £t a time.

(a) Write an expression for the number of pounds taken by the firm for n hirings at £t a time.

(b) Write a formula for the profit £P made by the company, giving P in terms of C, n and t.

(c) Use your formula to find the profit made from 35 hirings at £3 a time if the company's daily running costs total £60.

(d) Explain what happens if there are only 8 hirings at £3 and running costs still total £60.

(e) If the running costs remain at £60 per day and the cost of hiring a boat stays at £3, copy and complete the following table which compares the profit, £P, with the number of hirings, t:

Number of hirings, t	0	10	25	50	60	65
Profit, P(pounds)	⁻60			90		

(f) Draw a pair of axes as shown.
The P-axis should range from ⁻60 to 140. (Use a scale of 1 cm to £10.) The t-axis should range from 0 to 70. (Use a scale of 1 cm to 5 hirings.)
Draw a graph to show how P varies with t.

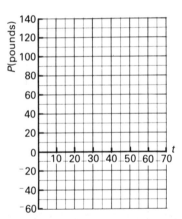

(g) Use your graph to find the number of hirings needed to make a profit of £66.

(h) Explain what the negative values on the profit axis mean.

(i) Use your graph to find the number of hirings needed for the company to break even (i.e. when the money received by the company is the same as the money paid out by the company).

(j) Write a formula for P in terms of t.

(k) Rewrite the formula to give t in terms of P.

4. Two identical triangles with sides measuring x cm, y cm and z cm are placed together in six different ways as shown below:

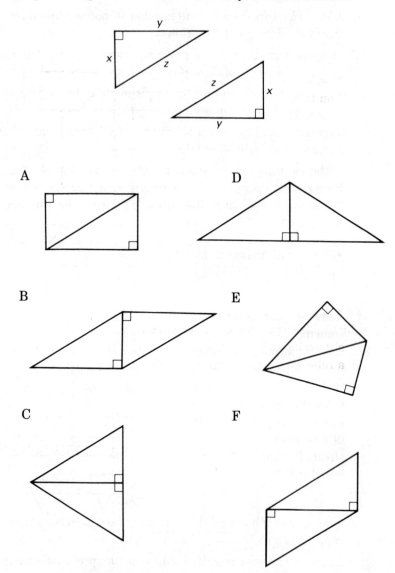

(a) Write the name of each shape formed.
(b) For each different arrangement, write an expression for its perimeter in terms of x, y and z.
(c) If $z > y > x$, which two arrangements have the largest perimeters?

Revision Exercise XVIII

1. Describe complete-
ly the transforma-
tion that would
map:

(a) D on to F

(b) F on to E

(c) D on to E

(d) D on to C

(e) C on to E

(f) A on to F

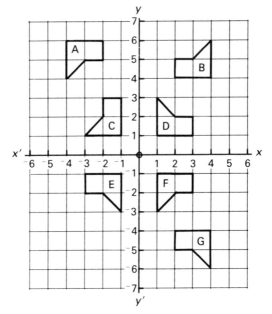

2. Which shape is an image of G under:

(a) a translation?

(b) a reflection?

3. (a) Copy the pair of
axes shown and
draw the given
parallelogram.
Label it P.

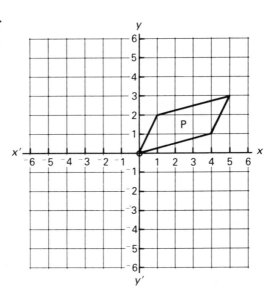

(b) Reflect the parallelogram in the y-axis. Draw the image and label it Y.

(c) Rotate parallelogram P through 90° clockwise and label the image, R.

(d) Rotate parallelogram R through 90° clockwise and label this image S.

(e) Describe completely the single transformation that would map shape Y on to shape S.

4. Answer the following without plotting any points. Find the co-ordinates of the image of each point if:

(a) (3, ⁻5) is reflected in the y-axis.

(b) (4, ⁻1) is reflected in the x-axis.

(c) (⁻2, ⁻3) is translated 2 units to the right (parallel to the x-axis) and 3 units downwards (parallel to the y-axis).

(d) (6, ⁻2) is rotated through 180° about the origin.

(e) (⁻3, 7) is rotated 90° anticlockwise about the origin.

5. Copy the following shapes then reflect each in line m:

(a)

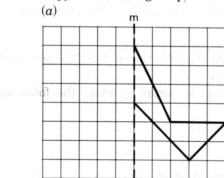

(b)

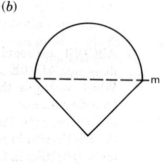

6. (a) Copy the pair of axes shown opposite. (Use a scale of 1 cm to 1 unit.)

(b) Draw the given triangle and label it ABC.

(c) Draw the line $y = 6$.

(d) Draw the reflection of $\triangle ABC$ in line $y = 6$ and label the image $A_1B_1C_1$.

(e) Draw the reflection of A'B'C' in the x-axis and label the image $A_2B_2C_2$.

(f) $\triangle ABC$ is labelled clockwise. Write what you notice about the labelling of its first reflection, $\triangle A_1B_1C_1$ and its second reflection, $A_2B_2C_2$.

(g) Draw the straight line $x = {}^{-}1$.

(h) Reflect $\triangle ABC$ in $x = {}^{-}1$ and label its image $A_3B_3C_3$.

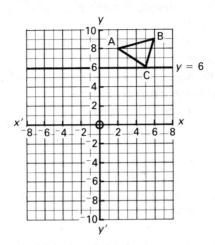

7. (a) If $({}^{-}2, 4)$ was reflected in the line $x = 3$, what would be the co-ordinates of its image?

(b) If $({}^{-}3, {}^{-}1)$ was reflected in the line $y = {}^{-}2$, what would be the co-ordinates of its image?

8. ABCDEF is a regular hexagon. O is the point of intersection of diagonals AD, BE and CF.

Which vertex is the image of vertex D under the following transformations:

(a) a rotation of 180° about O,

(b) a reflection in BE,

(c) a reflection in CF,

(d) a rotation through 120° clockwise about O,

(e) a reflection in AD?

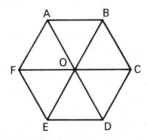

9. In hexagon ABCDEF, if F is the image of B following a clockwise rotation, through how many degrees is the rotation?

10. (a) Copy the straight line AB on to squared paper and mark points P and Q.

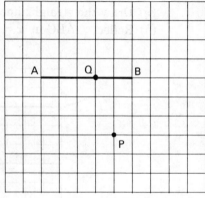

(b) Rotate AB through 180° about P. Label the image, A_1B_1.

(c) Rotate AB through 90° clockwise about P. Label the image A_2B_2.

(d) Rotate AB anticlockwise through 90° about P. Label the image A_3B_3.

(e) Rotate AB anticlockwise through 90° about Q. Label the image A_4B_4.

11. (a) Draw a pair of axes as shown then draw triangle ABC.

(b) Draw the image of △ABC if △ABC is rotated through 180° about C.

(c) Draw the image of △ABC when △ABC is rotated through 180° about the mid-point of BC.

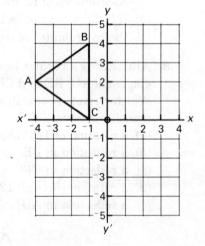

12. Draw an enlargement of the given shape so that each side is twice as long (that is, the scale factor is 2).

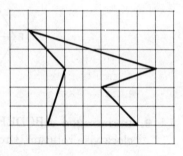

314

13. Copy the triangles (one is an enlargement of the other).

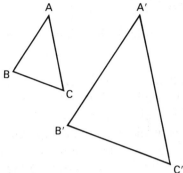

(*a*) Find the centre of enlargement.
(*b*) What is the scale factor of this enlargement?

14. L'M'N' is an enlargement of LMN.

(*a*) Find the co-ordinates of the centre of enlargement.
(*b*) What is the scale factor of the enlargement?

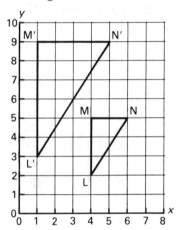

Revision Exercise XIX

1. Two circles have diameters of 10 cm and 15 cm. Find in their simplest form:
   (*a*) the ratio of their radii,
   (*b*) the ratio of their circumferences.

2. Which of the following are in the ratio 3 to 4?
   A £3 to 40 p   D 90° to 120°
   B 15 mm to 2 cm   E £1.20 to £1.60
   C 4 km to 3 km   F 3 cm to 4 m

3. In an exam, the ratio of the number of passes to the number of failures was 7 : 3. If there were 21 passes, how many failed?

4. A model is 7.5 cm long. If the scale of the model is 1 to 32, find the true length in metres.

5. A bus travelled 168 miles in $3\frac{1}{2}$ h. Calculate its average speed in miles per hour.

6. By taking 20 equal-length paces, Hugh walked 15 m.
    (*a*) How far would he walk in taking 70 paces?
    (*b*) How far would he walk in taking 45 paces?
    (*c*) How far would he walk in taking 300 paces?
    (*d*) How many paces must he take to walk 180 m?
    (*e*) How many paces must he take to walk 125 m?

Revision Exercise XX

1. Write whether or not the given pairs of shapes are similar:
    (*a*)

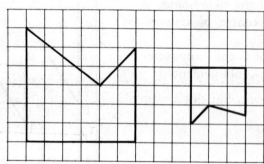

    (*b*)

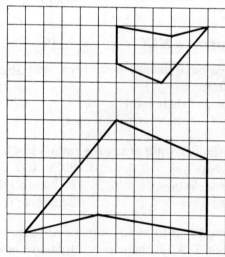

2. Write whether or not the given pairs of triangles are similar:

(*a*)   (*b*)

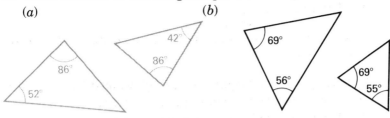

3. Write whether or not the given pairs of triangles are similar:

(*a*)   (*b*)

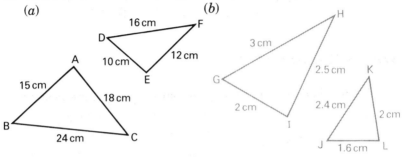

4. Write whether or not the given pairs of triangles are similar. (The marked angles are equal.)

(*a*)   (*b*)

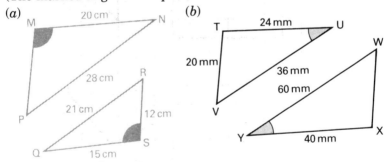

5. Triangles PQR and PTS are similar. PS = 12 cm, SR = 3 cm and ST = 6 cm. Calculate the length of RQ.

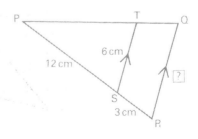

6. Triangles ABC and EDC are similar. DE = 12 cm, CE = 9 cm, AC = 12 cm and BC = 15 cm. Calculate the length of BA.

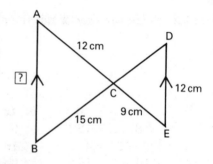

7. Alex is 1.5 m tall. How far is she from a 28 m tall beech tree?

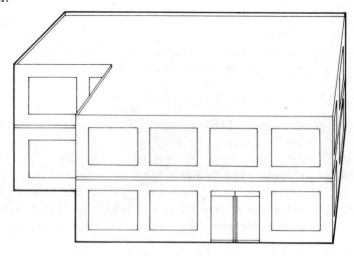

8. Some pupils made a model of their school. They made it 16 cm high using a scale of 1 to 40. What is the actual height of the school?

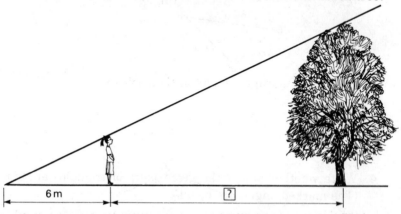

Revision Exercise XXI

A Solve these equations:

1. (a) $3x + 7 = 19$ (b) $6x - 5 = 13$ (c) $2x + 4 = 16$
2. (a) $2x - 7 = 8$ (b) $2x + 4 = 9$ (c) $3x - 2 = 9$
3. (a) $3x + 5 = {}^-4$ (b) $4x - 5 = {}^-13$ (c) $2x + 14 = 4$
4. (a) $18 - x = 12$ (b) $9 - 2x = 3$ (c) $24 - 3x = 9$
5. (a) $9 - 2x = 4$ (b) $16 - 4x = 10$ (c) $20 - 5x = 7$
6. (a) $10 - x = 15$ (b) $9 - 3x = 15$ (c) $18 - 2x = 32$
7. (a) $4(x - 3) = 16$ (b) $3(2x + 5) = 21$
8. (a) $3x - 7 = x + 2$ (b) $6x + 5 = 2x + 17$
9. (a) $10 + 2(x - 3) = 16$ (b) $2(3x - 4) = 3x + 7$
10. (a) $5(3x - 1) = 2(6x + 2)$ (b) $9 - 2(x - 4) = 3(2x - 5)$

B For each question, form an equation then solve it:

1. Find the value of x:

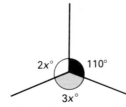

2. When 6 is added to a number, the answer is 19. Find the number.

3. Marcia thought of a number, doubled it then added 9. The answer was 25. What number did Marcia think of?

C 1. $x - 2 \leqslant 4$:
    (a) Give two possible values of x that are whole numbers.
    (b) Give one non-integral value of x.

2. Give two values of x such that $x + 9 > 14$.

3. Give two possible values of x such that $2x \geqslant 12$ and $x < 7$.

Revision Exercise XXII

1. Draw a pair of axes where x ranges from $^-4$ to 4 (use a scale of 1 cm to 1 unit) and y ranges from $^-15$ to 15 (use a scale of 2 cm to 5 units).

(*a*) Draw the graph of $y = 7$.

(*b*) On the same pair of axes, draw the graph of $y = 3x + 1$.

(*c*) Write the co-ordinates of the point of intersection of the two graphs.

(*d*) If the graph of $y = 3x + 1$ crosses the y-axis at A, what are the co-ordinates of A?

2. Draw a pair of axes where x ranges from 0 to 8 $(0 \leqslant x \leqslant 8)$ and y ranges from $^-6$ to 8 $(^-6 \leqslant y \leqslant 8)$. Use a scale of 1 cm to 1 unit on both axes.

(*a*) Draw the graph of $x + y = 7$.

(*b*) On the same pair of axes, draw the graph of $y = x - 5$.

(*c*) Write the co-ordinates of the point of intersection of the two graphs.

3. The graphs of $y = x - 6$ and $y = 6 - x$ are shown in the diagram (one is labelled D and the other U). Which is which?

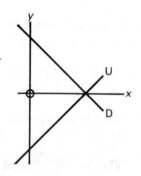

4. The diagram shows a sketch of the graph of $x + 2y = 9$. It crosses the x-axis at A and the y-axis at B. What are the co-ordinates of:

(*a*) A?   (*b*) B?

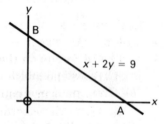

5. Draw a pair of axes where x ranges from $^-3$ to 5 and y from $^-4$ to 10. (Use a scale of 1 cm to 1 unit on both axes.) Draw the graph of $y = 3(x + 1)/2$.

(*a*) At which point does the graph cross the y-axis?

(*b*) At which point does the graph cross the x-axis?

6. The graph of $y = 2x + c$ passes through the point (4, 12). Find the value of c.

7. Draw a pair of axes where x ranges from 0 to 7 and y from $^-9$ to 5. (Use a scale of 1 cm to 1 unit.) Draw the graph of $y = 2x - 9$ and use it to solve the equation $2x - 9 = 0$.

8. For each graph, find the gradient of line segment PQ:

(a) (b)

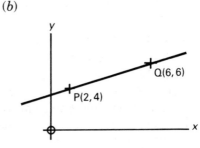

9. Find the gradient of the straight-line segment joining the points A($^-2$, $^-1$) and B(1, 5).

10. What is the gradient of each of the following graphs?
(a) $y = 2x - 7$ (b) $y = 5x + 1$ (c) $y = \frac{1}{2}x + 6$

11. What is the gradient of each of the following graphs?
(a) $y = 2 - 3x$ (b) $y = {^-2}x + 9$ (c) $y = {^-5} + 2x$

12. What is the gradient of each of the following graphs?
(a) $3y = 9x - 12$ (b) $2y = 14 - 6y$ (c) $2x + y = 15$

13. Draw a pair of axes where x ranges from $^-5$ to 5 and y from 0 to 25. Use a scale of 1 cm to 1 unit on the x-axis and 2 cm to 5 units on the y-axis.
(a) Draw the graph of $y = x^2$.
(b) On the same pair of axes, draw the graph of $y = 4$.
(c) Write the co-ordinates of the points of intersection of the two graphs.

23 Bearings and Scale Drawings

The diagram shows the positions of two ships, one at A and the other at B.

B is at a *bearing* of 060° from A.

The bearing is the angle measured clockwise from *North*.

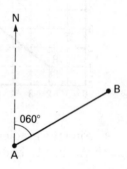

Note that if a bearing is measured from A, the angle itself must be at A. The bearing can measure anything up to 360° and must be written using three figures. It is called a three-figure bearing. (It is also sometimes called a circular bearing or an absolute bearing.)

Note   To write that B is at a bearing of 60° from A (for the diagram above) is incorrect. (Three digits must be used.)

Exercise 1

For each diagram, give the bearing of B from A:

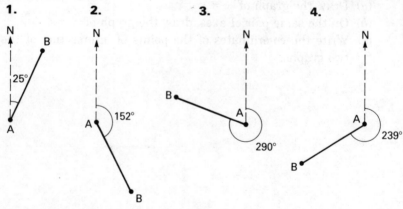

1. **2.** **3.** **4.**

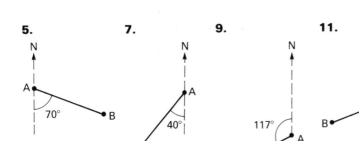

5.

7.

9.

11.

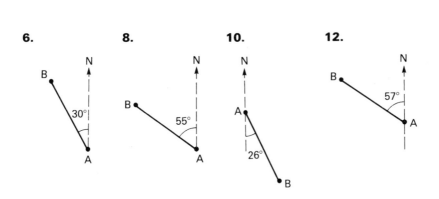

6.

8.

10.

12.

Exercise 2

1. What is the bearing of:
   (a) P from O?
   (b) Q from O?

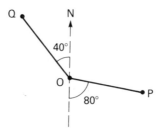

2. What is the bearing of:
   (a) T from X?
   (b) U from X?

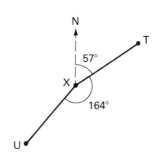

323

3. What is the bearing of:
   (*a*) A from P?
   (*b*) B from P?

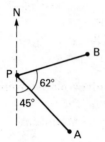

6. What is the bearing of:
   (*a*) X from M?
   (*b*) Y from M?

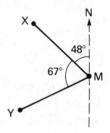

4. What is the bearing of:
   (*a*) D from C?
   (*b*) F from C?

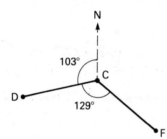

7. What is the bearing of:
   (*a*) G from B?
   (*b*) H from B?

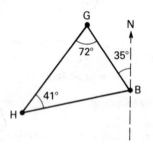

5. What is the bearing of:
   (*a*) J from K?
   (*b*) L from K?

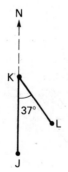

8. What is the bearing of:
   (*a*) R from T?
   (*b*) V from T?

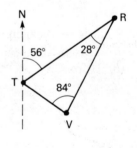

Exercise 3

Use the map below for these questions.

1. From High Tor, which places are at these bearings?
   (a) 036°
   (b) 156°
   (c) 320°
   (d) 205°
   (e) 283°

2. Give the distance and bearing of the given places from High Tor:
   (a) Crystal Lake
   (b) Copper Cliffs
   (c) Wide Plains
   (d) Snow Mountains

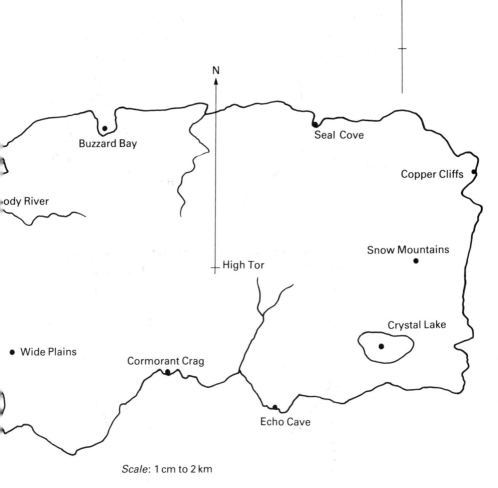

Scale: 1 cm to 2 km

Exercise 4

A Find the size of the smallest angle between the given bearings:

1. 165° and 085°   **5.** 270° and 046°

2. 215° and 170°   **6.** 149° and 203°

3. 304° and 225°   **7.** 092° and 261°

4. 040° and 310°   **8.** 098° and 297°

B 1. A ship is sailing on a bearing of 074°. Find its new bearing if it turns through:

(*a*) 65° to starboard (the right), (*d*) 97° to port,

(*b*) 25° to port (the left),   (*e*) 123° to port,

(*c*) 97° to starboard,   (*f*) 180° to starboard.

2. An aeroplane is flying on a bearing of 287°.
Find its new bearing if it turns through:

(*a*) 40° to starboard,   (*d*) 108° to starboard,

(*b*) 55° to port,   (*e*) 180° to port,

(*c*) 108° to port,   (*f*) 218° to starboard.

Exercise 5

For each question, give the new bearing if you face the given direction then make an about-turn to face the opposite direction.

1. 060°   **3.** 035°   **5.** 285°   **7.** 019°   **9.** 301°

2. 110°   **4.** 125°   **6.** 214°   **8.** 326°   **10.** 182°

In the diagram, the bearing of P from Q is 075°. The bearing of Q from P is then called the *back bearing** (or *reverse bearing*). In the diagram, the back bearing is 255°.

*See Appendix 3, p.475.

326

Exercise 6

A The bearing of P from Q is given. Find the bearing of Q from P.

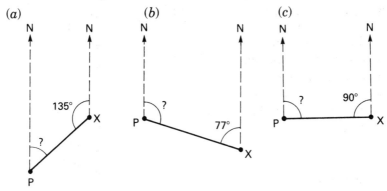

1. 110°   **2.**   **3.** 258°

4. 076°   **5.** 099°   **6.** 196°   **7.** 270°   **8.** 312°

B 1. A ship is sailing on a bearing of 125°. Another ship is sailing in exactly the opposite direction. On what bearing is this second ship sailing?

2. A hiker walked on a bearing of 028°. What would be the bearing of the return journey?

3. Two ships pass each other travelling in opposite directions. If one ship is travelling on a bearing of 283°, in what direction is the other ship sailing?

Exercise 7

1. Find the bearing of X from P:

(a)   135°   (b)   77°   (c)   90°

2. Find the bearing of:
  (a) B from A,
  (b) C from A,
  (c) C from B,
  (d) A from B,
  (e) A from C,
  (f) B from C.

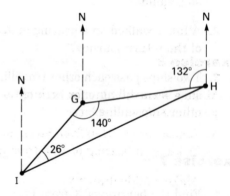

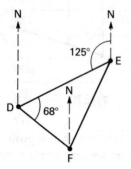

3. Find the bearing of:
  (a) D from E,
  (b) E from D,
  (c) F from D,
  (d) D from F.

4. Find the bearing of:
  (a) G from H,
  (b) H from G,
  (c) I from G,
  (d) G from I,
  (e) I from H,
  (f) H from I.

Exercise 8

Mark a point P near the centre of a piece of exercise paper then mark a North line.

Using a scale of 1 cm to 1 km, plot these points:

	Place	Distance from P (km)	Bearing
1.	A	8.0	055°
2.	B	6.5	132°
3.	C	4.7	165°
4.	D	6.5	230°
5.	E	9.3	320°
6.	F	5.6	270°
7.	G	4.4	090°
8.	H	2.9	180°
9.	I	7.2	028°
10.	J	5.1	293°

Exercise 9

Make a scale drawing for each of the following. Note that distances given are direct distances.

1. Doncaster is 70 km from Stockport on a bearing of 080°. Derby is 70 km at a bearing of 140° from Stockport.

Make a scale drawing using a scale of 1 cm to 10 km (1 : 1 000 000).
Find:

(a) the direct distance of Doncaster from Derby,

(b) the bearing of Doncaster from Derby,

(c) the bearing of Derby from Doncaster.

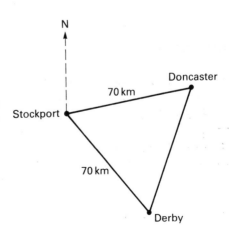

329

2. Yeovil is 55 km due south of Bristol. Salisbury is at a bearing of 124° from Bristol and 076° from Yeovil.

Make a scale drawing using a scale of 1 cm to 10 km. Find:

(a) the bearing of Bristol from Salisbury,

(b) the bearing of Yeovil from Salisbury,

(c) the distance of Salisbury from Bristol,

(d) the distance of Yeovil from Salisbury.

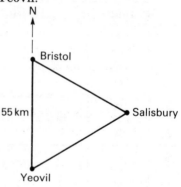

3. (a) Hexham is 50 km from Brough on a bearing of 015°. Measure the distance on the given diagram and write the scale used in the form 1 cm to ⟨?⟩ km.

(b) Copy the diagram and mark the position of Durham if it is 55 km from Brough at a bearing of 059°.

(c) How far is Hexham from Durham?

Exercise 10

1. A ship sailed 24 km on a bearing of 035°, sailing from X to Y. It then changed course to a bearing of 110° and sailed 16 km to Z.

Make a scale drawing using the scale of 1 cm to represent 4 km.

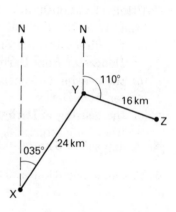

(*a*) Measure length XZ in centimetres giving its length correct to one decimal place.

(*b*) How many kilometres is Z from X.

(*c*) Join XZ and find angle YXZ.

(*d*) What is the bearing of Z from X?

(*e*) What is the bearing of X from Z?

2. A ship sailed 56 km on a bearing of 130° sailing from P to Q. It then changed course and sailed from Q to R, a distance of 76 km on a bearing of 275°. Make a scale drawing. Use the scale of 1 cm to represent 8 km. Join PR.

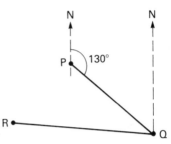

(*a*) Measure PR in centimetres correct to one decimal place (that is, to the nearest millimetre).

(*b*) Give the distance from P to R in kilometres.

(*c*) Measure ∠QPR.

(*d*) Give the bearing of R from P.

(*e*) Give the bearing of P from R.

Exercise 11

Make a scale drawing for each of the following.
By measuring, find for each question:

(*a*) the direct distance between the starting point and finishing point,

(*b*) the bearing of the finishing point from the starting point,

(*c*) the bearing of the starting point from the finishing point.

1. A ship sailed 4 km due east, followed by 4 km due north.

2. A ship sailed 4 km at 270°, followed by 3 km at 000°.

3. A ship sailed 6 km at 090°, followed by 5 km at 045°.

4. A ship sailed 6 km at 090°, followed by 5 km at 310°.

5. A ship sailed 3 km at 045°, followed by 8 km at 135°.

6. A ship sailed 35 km at 270°, followed by 25 km at 180°.

7. A ship sailed 80 km at 340°, followed by 25 km at 150°.

8. A ship sailed 2.8 km at 239°, followed by 4.2 km at 146°.

9. A ship sailed 3 km at 000°, then 10 km at 090°, then 6 km at 180°.

10. A ship sailed 5 km at 135°, then 9 km at 045°, then 8 km at 180°.

11. A ship sailed 63 km at 312°, then 84 km at 108°, then 46 km at 000°.

12. A ship sailed 150 km at 336°, then 150km at 198°, then 52 km at 137°.

Exercise 12

1. An aeroplane flew from airport A for 3200 km on a bearing of 035° to airport B. It then flew to airport C, a distance of 2000 km on a bearing of 140° from B.
   (a) Using a scale of 1 cm to 400 km, make a scale drawing of the route and use it to find the distance and bearing of airport C from airport A.
   (b) How long would the aeroplane take to fly from A to B to C at 800 km/h?
   (c) At what bearing should the aeroplane fly to return directly from C to A?

2. A hiker walked 2 km on a bearing of 325° to reach point P. She then walked to Q, 3 km from P and at a bearing of 050° from P.
   (a) Make a scale drawing using a scale of 1 cm to 0.5 km (2 cm to 1 km) and use it to find the distance and bearing of Q from the starting point.
   (b) At what bearing should the hiker walk from Q to return to the starting point by a direct route?
   (c) Walking at a steady 4 km/h, how long would the hiker take to complete the whole journey from the start to P then to Q then returning straight to the same starting point.

332

3. Ship A is 50 nautical miles due east of position P. Ship B is 65 nautical miles and on a bearing of 330° from P, while ship C is 40 nautical miles and on a bearing of 215° from P.
   (a) Make a scale drawing to show all the positions. Use a scale of 1 cm to 10 nautical miles.
   (b) Give the distance and bearing of B from A.
   (c) Give the distance and bearing of C from A.
   (d) Give the distance and bearing of C from B.
   (e) Give the bearing of A from B.
   (f) Give the bearing of A from C.
   (g) Give the bearing of B from C

Exercise 13   **M**

1. Use a copy of the map on p.325.

2. Start at High Tor and plot the following route:

   10 km at 065°
   7 km due west
   14 km at 222°
   6 km at 345°

3. How far is the finishing point from High Tor?

4. If you followed the instructions in a different order, would you arrive at the same finishing point?

Exercise 14

1. From a point 30 m from the foot of a building, the angle of elevation of the top of the building is 20°. Make a scale drawing using a scale of 1 cm to 5 m and find the height of the building.

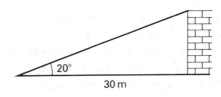

20°

30 m

2. From the top of a building 20 m high, the angle of depression of a point on the ground is 24°. Make a scale drawing using a scale of 1 cm to 5 m and use it to find the distance of the point from the foot of the building.

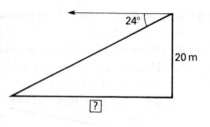

3. A stick of length 2.7 m casts a 2.1 m shadow. Make a scale drawing using a scale of 1 cm to 30 cm. Find the angle of elevation of the Sun.

4. A building is 20 m high. From a position on the ground the angle of elevation of the top of the building is 25°.
Make a scale drawing using a scale of 1 cm to 5 m, and use it to find the distance of that point from the foot of the building.

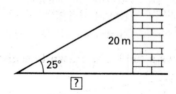

5. A boy measured the angle of elevation of the top of a building to be 18°. If the measurement was taken at eye level from a point 40 m from the foot of the building, and if the boy's height was 1.5 m up to eye level, make a scale drawing to help you to find the height of the building. Use a scale of 1 cm to 4 m.

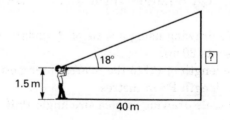

1. The diagram, not drawn to scale, is of a field.

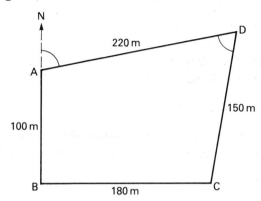

Side AB runs from North to South and is 100 m long. Side BC runs from West to East and is 180 m long. D is 220 m from A and 150 m from C.

Make a scale drawing. Use a scale of 1 cm to represent 20 m.

(a) Find the bearing of D from A.

(b) Find the bearing of A from D.

(c) Measure diagonal AC. Give the answer to the nearest metre.

(d) Measure AD̂C.

2. The diagram shows a field, not drawn to scale.

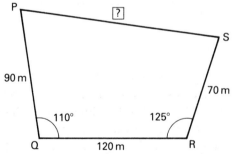

PQ = 90 m,　QR = 120 m,　PQ̂R = 110°,　QR̂S = 125°　and RS = 70 m.

Make a scale drawing using a scale of 1 : 2000 (1 cm represents 20 m).

(a) Find the length of PS to the nearest millimetre.

(b) Give the length PS in metres.

(c) Use the scale drawing to measure angle PSR.

Here are three views of a garden shed:

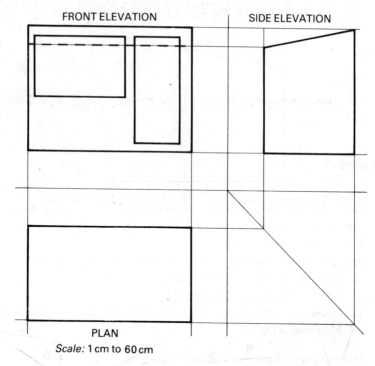

This method of drawing is called *orthographic projection* and is used in technical drawing.

1. How long is the shed?

2. What is the width of the shed?

3. Give the dimensions of the door.

4. Give the dimensions of the window.

5. The roof of the shed slopes downwards from the front to the back. What is the height of the shed:
(*a*) at the front? (*b*) at the back?

24 Congruency

Two figures are *congruent* if they have the *same shape* and the *same size*.

Exercise 1

For each question, write which two figures are congruent:

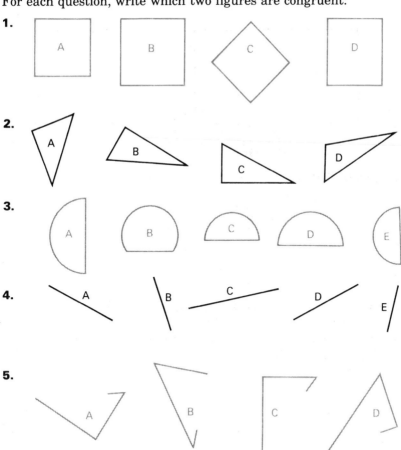

1. A   B   C   D

2. A   B   C   D

3. A   B   C   D   E

4. A   B   C   D   E

5. A   B   C   D

6.

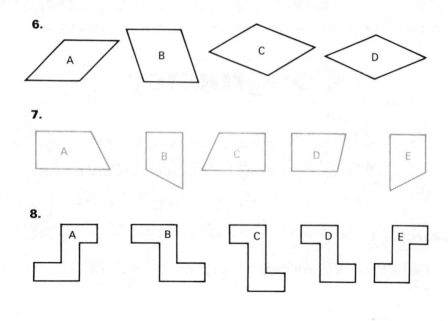

7.

8.

Exercise 2

Where possible, construct the following triangles and answer the questions given. If it is possible to construct more than one triangle from the information given, then construct more than one and state how many different triangles can be constructed.

1. (Three sides given, which can be abbreviated to SSS.)
   (a) Construct $\triangle ABC$: $AB = 48$ mm, $BC = 34$ mm, $CA = 29$ mm. Measure $\hat{A}$.
   (b) Construct $\triangle XYZ$: $XY = 76$ mm, $YZ = 38$ mm, $ZX = 32$ mm.

2. (a) (Two sides and the included angle given—that is, the angle between the two given sides, which can be abbreviated to SAS.)
   Construct $\triangle DEF$: $EF = 61$ mm, $DE = 52$ mm, $\angle DEF = 34°$. Measure DF.
   (b) (Two sides and one angle—not the included angle, which can be abbreviated to ASS.)
   Construct $\triangle JKL$: $KL = 68$ mm, $LJ = 52$ mm, $J\hat{K}L = 36°$. Measure JK.

3. (Two angles and one side given, which can be abbreviated to AAS.)
   (*a*) Construct △LMN: LM = 50 mm,   ∠L = 39°,   ∠M = 52°.
   Measure LN.
   (*b*) Construct △LMN: LM = 50 mm,   ∠L = 39°,   ∠N = 89°.
   Measure LN.

4. (Three angles given, which can be abbreviated to AAA.)
   (*a*) Construct △UVW: $\hat{U}$ = 84°, $\hat{V}$ = 46°, $\hat{W}$ = 50°.
   Measure UV.
   (*b*) Construct △PQR: $\hat{P}$ = 71°, $\hat{Q}$ = 48°, $\hat{R}$ = 51°.

5. (Right-angle, hypotenuse and one side given, which can be abbreviated to RHS.)
   Construct △QRS: $\hat{R}$ = 90°, RS = 42 mm, QS = 103 mm.
   Measure QR.

Exercise 3

Draw as many different triangles as possible with one side measuring 64 mm and where two of the angles are 42° and 79°. How many different possible triangles are there?

Congruent Triangles (notes and information)

1. Three sides (SSS)

A triangle can be constructed from three given sides as long as the sum of the two shortest sides is greater than the longest side.
Given three sides, exactly one triangle can be constructed. This is shown in Exercise 2, question 1(*a*), opposite, where only one triangle can be constructed from the three given sides. (SSS is a congruency.)

This congruency is used in triangulation in surveying.

From two positions that have already been drawn (A and B in the diagram), a third position can always be found by using the distances to that point from each of the two known points.

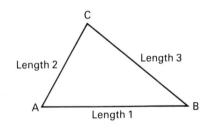

339

2. Two sides and one angle:

(a) *Two Sides and the Included Angle (SAS)*

SAS is a congruency. In Exercise 2, question 2(a), p.338, only one triangle could be constructed from the given information.
This is also explained using the diagram.
If AB is one of the given sides and is drawn as shown, then the given angle can be measured (angle A in the diagram). Side 2 can then be drawn from A and since it is a fixed length it must end at a fixed point. Only the one triangle is possible.

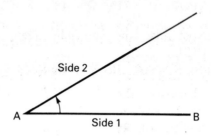

(b) *Two Sides and an Angle other than the Included Angle (ASS)*
This is not a congruency.

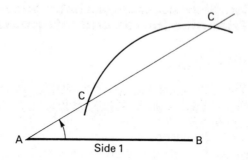

If AB is one of the given sides and is drawn as shown, then the given angle can be measured (Angle A in the diagram).
Since the angle is not the included angle, the other given side cannot be AC and must be BC.
However, as the diagram shows, there are two possible positions for C (check that the length BC is the same for both positions of C). Hence, two different triangles can be constructed from the same information, so ASS is not a congruency.

3. Two Angles and One Side (AAS)

Exercise 2, question 3, p.339, shows that given two angles and one side, only one triangle is possible. Note that the position of the side with respect to the angles must be given. Exercise 3, p.339, shows that if the position of the side is not given then three different triangles are possible. Given two angles and one side, several people would all draw triangles of the same shape and size as long as they all positioned the side in the same place. The side is normally therefore referred to as a *corresponding* side. Hence, AAcorr.S is a congruency.

This congruency is also used in surveying (in what is normally referred to as a double-station survey) and in navigation.

(*a*) The angles marked in the diagram can be found. Since the distance between the landmarks would be known, the two angles and the side fix the position of the ship.

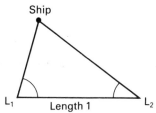

(*b*) From a ship or an aircraft the bearings of two landmarks can be found. Since the distance between the landmarks is known, the two angles and the side once again fix the position of the ship.

4. Three Angles (AAA)

This condition gives similar triangles. The triangles obtained can be any size, so AAA is not a congruency. This is shown in Exercise 2, question 4, p.339.

5. Right-angle, Hypotenuse and One Side (RHS)

RHS is a congruency. Exercise 2, question 5, p.339, shows that only one triangle is possible.

The four congruencies are: SSS, SAS, AAcorr.S, RHS

The symbol '≡' means 'is congruent to'
so '△ABC ≡ △PQR' reads '△ABC is congruent to △PQR'.

Exercise 4

For each question, write whether or not the given pair of triangles is congruent. Where possible, state the congruency.

(*Note* The triangles have been drawn to look different, but they may in fact be congruent—it is the measurements or the equality markings that matter, not the general appearance.)

A 1.

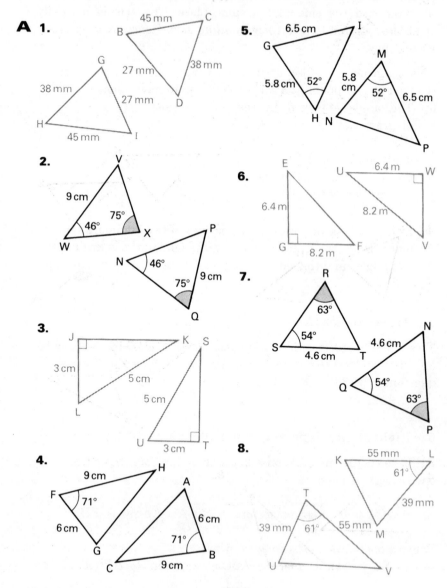

9.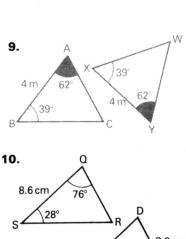

10.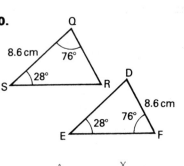

11.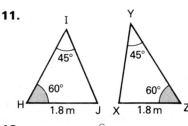

12.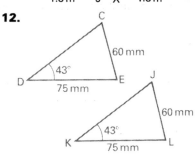

B 1.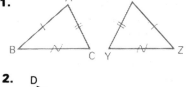

2.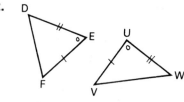

3.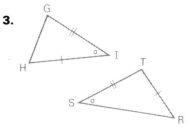

4.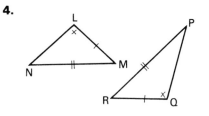

5.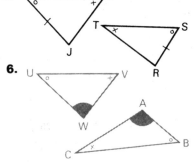

6.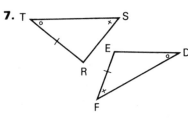

7.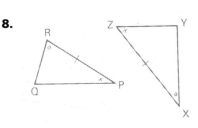

8.

9.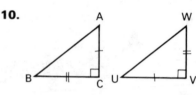

10.

Exercise 5 M

Copy each pair of congruent triangles from Exercise 4, part B. Mark *all* the equal angles and *all* the equal sides on each pair of congruent triangles.

Exercise 6 Constructions Based on Congruency

1. A base line in a triangulation survey runs North to South and is 30 m long. A building is East of this base line. If one corner of the building is 25 m from the northern end of the base line and 34 m from the southern end, make a scale drawing. By measuring, find the bearing of the corner of the building from both ends of the base line.

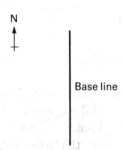

2. Three people, Audrey, Bimal and Clive stand so that Bimal is 50 m from Audrey at a bearing of 038°. Clive is 65 m from Audrey at a bearing of 098°. Make a scale drawing using a scale of 1 cm to 10 m.
   (*a*) How far is Clive from Bimal?
   (*b*) What is the bearing of Clive from Bimal?
   (*c*) What is the bearing of Bimal from Clive?

3. Two lighthouses are 20 km apart, one being due north of the other. A ship obtains a fix from these lighthouses and is at a bearing of 019° and 135° from them. Make a scale drawing and use it to find the distance of the ship from the lighthouses.

4. A and B are 100 km apart and B is due east of A. A is at a bearing of 330° from an aeroplane. B is at a bearing of 047° from the same aeroplane. Make a scale drawing and use it to find the distance of the aeroplane from A and from B.

344

25 Pythagoras

For any right-angled triangle:
the area of the square on the hypotenuse is equal to the sum of the areas of the squares on the other two sides.

$b^2 = a^2 + c^2$

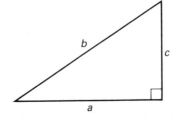

Exercise 1

1. △ABC is right-angled at B.
AB = 3.6 cm, BC = 4.8 cm.
Find:
(a) the area of the square on side AB,
(b) the area of the square on side BC,
(c) the area of the square on side AC,
(d) the length of AC.

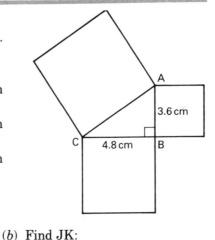

2. (a) Find XY: (b) Find JK:

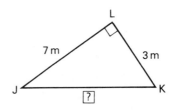

3. (*a*) Find PQ: (*b*) Find EG:

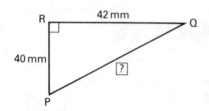 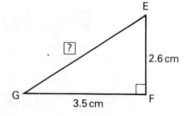

4. In right-angled triangle DEF, find DF to 3 s.f., if:
   (*a*) DE = 5 cm and EF = 9 cm
   (*b*) DE = 25 mm and EF = 65 mm
   (*c*) DE = 0.3 m and EF = 0.8 m
   (*d*) DE = 3.4 cm and EF = 5.3 cm
   (*e*) DE = 1.2 m and EF = 1.9 m

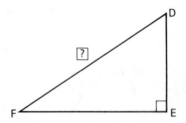

5. Mr Newman wants a piece of cord to stretch across a diagonal of a rectangular plot of garden. If the plot measures 11 m by 7 m, how long must the cord be?

6. Mrs Vincent's bookshelves are 2 m high and 0.9 m wide. To make them rigid she needs a piece of wood to reach from corner to corner. What length of wood is needed?

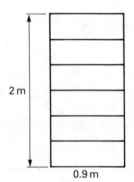

7. What length of support is needed for a shelf as shown in the diagram? The shelf is 21 cm wide and the support is to be fixed to the wall at a point 18 cm below the shelf. The other end of the support is 6 cm from the end of the shelf.

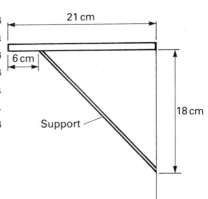

8. Mr Barker wanted to paint the wood under the guttering of his house. The guttering was 5.2 m above the ground. His ladder reached the guttering when its foot was placed against his neighbour's wall 3 m away. What length of ladder was it?

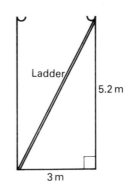

Exercise 2

1. △ABC is right-angled at B. AC = 17 cm, AB = 8 cm.
Find:
(a) the area of the square on side AC,
(b) the area of the square on side AB,
(c) the area of the square on side BC,
(d) the length of side BC.

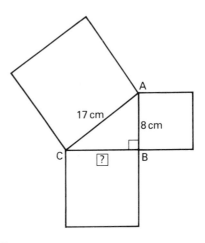

347

2. $\triangle PQR$ is right-angled at Q.
PR = 7.9 cm,   QR = 6 cm.
Find:

 (*a*) the area of the square on side PR,
 (*b*) the area of the square on side QR,
 (*c*) the area of the square on side PQ,
 (*d*) the length of side PQ.

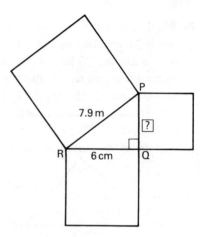

Note To find the area of square B (that is, the square on the hypotenuse) *add* the areas of the other two squares.

To find the area of one of the smaller squares, *sub-tract* the area of the other small square from the area of the large square on the hypotenuse.

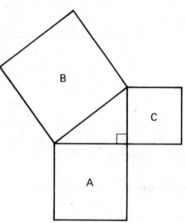

Exercise 3

For each triangle, find the missing side:

1.

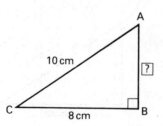

2.

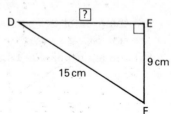

3.

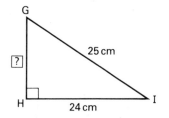

4.

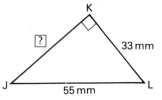

5.

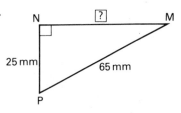

6.

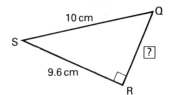

7.

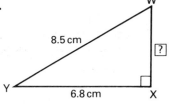

8.

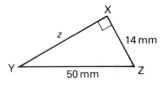

Exercise 4

1. (*a*) Find DE.

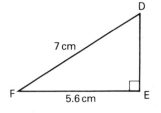

(*b*) Find *z*.

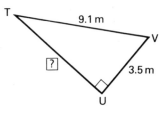

2. (*a*) Find the value of *j*.

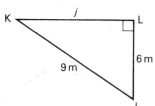

(*b*) Find TU.

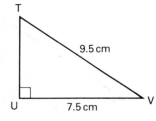

3. In right-angled triangle NPQ, find NP to three significant figures if:

   (a) NQ = 29 mm and PQ = 21 mm

   (b) NQ = 7.4 cm and PQ = 7 cm

   (c) NQ = 8 cm and PQ = 4.8 cm

   (d) NQ = 49 mm and PQ = 17 mm

   (e) NQ = 1.3 m and PQ = 0.4 m

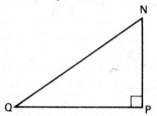

4. (a) Calculate the perpendicular height of an equilateral triangle with sides that measure 10 cm.

   (b) Calculate the area of the equilateral triangle.

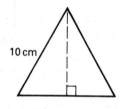

5. A chord of length 9 cm lies inside a circle of radius 5.3 cm. How far is the chord from the centre of the circle?

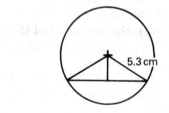

6. In the diagram, PS = 13 cm, PQ = 5 cm and RS = 9 cm.

   (a) How long is QR?

   (b) How long is PR?

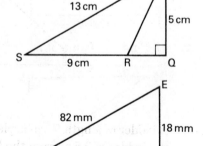

7. In right-angled triangle EFG, EG = 82 mm and EF = 18 mm. Calculate:

   (a) the length of GF,

   (b) the area of triangle EFG.

8. Using the information given in the diagram, calculate:
   (a) length AB,
   (b) length CD,
   (c) area of △ABD.

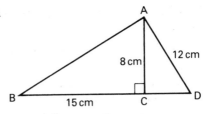

Exercise 5

1. A rectangular piece of card measures 15 cm by 10 cm.
   (a) Find its perimeter.
   (b) Find its area.

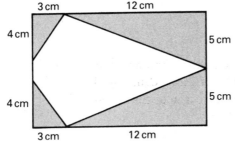

   The shaded triangles are cut out of the rectangular piece of card. Calculate:
   (c) the perimeter of the shape that is left,
   (d) the area of the shape that is left.

2. Calculate the height of the loft shown, using the dimensions given in the diagram.

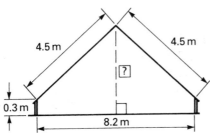

3. A ladder of length 5.5 m is placed against a wall so that the foot of the ladder is 2.5 m from the bottom of the wall. How high up the wall does the ladder reach?

4. A ladder of length 6.5 m is placed against a wall of a house so that it reaches 5.9 m up the wall. Without moving the foot of the ladder it is swung over to rest against the house next door. If it reaches 5.3 m up the wall of this house, calculate the distance between the two houses.

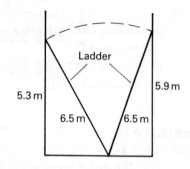

Exercise 6

1. Calculate the length of CD in the diagram given.

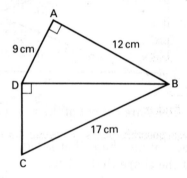

2. In the diagram, not drawn to scale, a vertical pole PQ, 8 m high, is on level ground AB. PB and WA are wires that hold the pole in position. AQ = BQ. Wire PB is 13 m long. W is 3.2 m below the top of the pole.

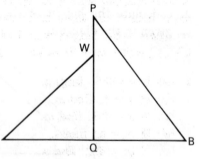

(*a*) Copy the diagram and show the various lengths.
(*b*) Calculate the length of BQ (to one decimal place).
(*c*) Calculate the length of WA (to one decimal place).

26 Simultaneous Equations

Exercise 1

A A man and his son practised a balancing act. The boy stood on the man's head and their total height was 10 ft. If we let the man's height be m ft and his son's height s ft, then $m + s = 10$.

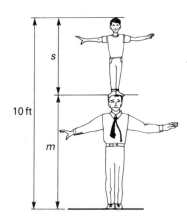

10 ft

1. (a) If $m = 6$, find s.
   (b) If $m = 6.2$, find s.
   (c) If $m = 5.7$, find s.
   (d) If $s = 3.5$, find m.
   (e) If $s = 5$, find m.
   (f) If $s = 3.6$, find m.

2. Write three more possible pairs of values for m and s.

B A man has two sons who are the same height. In their balancing act their total height was 14 ft. We can write $m + 2s = 14$.

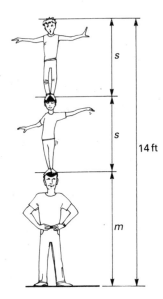

14 ft

1. (a) If $s = 3.5$, find m.
   (b) If $s = 4$, find m.
   (c) If $s = 3.6$, find m.
   (d) If $m = 6$, find s.
   (e) If $m = 6.2$, find s.
   (f) If $m = 5.7$, find s.

2. Write three more possible pairs of values for m and s.

353

C Which pair of answers to part B are the same as a pair of answers to part A?

D If the men in parts A and B have the same height, and if all three boys have the same height, then:

$$m + 2s = 14$$
and $$\quad m + s = 10$$

1. How tall are the boys? (The diagram may help.)

2. How tall are the men?

Note If $\qquad\qquad\qquad m + 2s = 14 \qquad$ [1]

and $\qquad\qquad\qquad m + s = 10 \qquad$ [2]

Equation [1] − Equation [2] gives $\qquad \underline{\underline{s = 4}}$

Substituting in [2] $\qquad\qquad\qquad\qquad m + s = 10$

so $\qquad\qquad\qquad\qquad\qquad\qquad\qquad m + 4 = 10$

∴ $\qquad\qquad\qquad\qquad\qquad\qquad\qquad \underline{\underline{m = 6}}$

so $m = 6$ and $s = 4$

that is, the men are 6 ft tall and the boys 4 ft tall.

The two equations have been solved *simultaneously*.

Exercise 2

Solve these pairs of simultaneous equations:

A **1.** $m + 2s = 16$
    $m + s = 11$

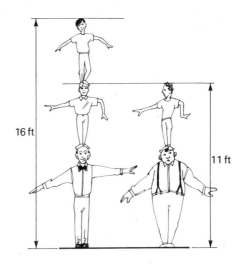

16 ft   11 ft

2. $m + 5s = 15$
    $m + 4s = 13$

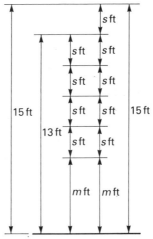

15 ft   13 ft   s ft   s ft   15 ft   s ft   s ft   s ft   s ft   s ft   s ft   s ft   s ft   m ft   m ft

B **1.** $m + 4s = 12$
    $m + 3s = 11$

2. $2m + 3s = 17$
    $2m + 2s = 14$

3. $3m + 2s = 22$
    $2m + 2s = 16$

4. $2m + 5s = 25$
    $m + 5s = 20$

5. $4m + s = 15$
    $m + s = 6$

6. $5m + 4s = 29$
    $3m + 4s = 19$

Exercise 3

Which pair of values satisfies both equations simultaneously?

1. $m + 3s = 19$
$m + s = 11$

- A. $m = 4,$ $s = 5$
- B. $m = 10,$ $s = 3$
- C. $m = 4,$ $s = 7$
- D. $m = 7,$ $s = 4$

5. $a + b = 8$
$a - b = 2$

- A. $a = 5,$ $b = 3$
- B. $a = 5\frac{1}{2},$ $b = 2\frac{1}{2}$
- C. $a = 6,$ $b = {}^-4$
- D. $a = 3,$ $b = 5$

2. $x + 6y = 27$
$x + y = 12$

- A. $x = 3,$ $y = 4$
- B. $x = 8,$ $y = 4$
- C. $x = 10,$ $y = 2$
- D. $x = 9,$ $y = 3$

6. $3g + h = 13$
$3g - 2h = 1$

- A. $g = 4,$ $h = 1$
- B. $g = 3.5,$ $h = 2.5$
- C. $g = 3,$ $h = {}^-4$
- D. $g = 3,$ $h = 4$

3. $4e + f = 24$
$4e + 5f = 40$

- A. $e = 4,$ $f = 8$
- B. $e = 5,$ $f = 4$
- C. $e = 5.5,$ $f = 2$
- D. $e = 3.5,$ $f = 10$

7. $2d + 3e = 4$
$4d + 3e = 14$

- A. $d = 8,$ $e = {}^-4$
- B. $d = 5,$ $e = {}^-2$
- C. $d = {}^-5,$ $e = 2$
- D. $d = {}^-8$ $e = 4$

4. $t + x = 8$
$3t + x = 13$

- A. $t = 3.5,$ $x = 4.5$
- B. $t = 2,$ $x = 6$
- C. $t = 2\frac{1}{2},$ $x = 5\frac{1}{2}$
- D. $t = 1\frac{1}{2},$ $x = 6\frac{1}{2}$

8. $2p + 4q = {}^-4$
$2p - 3q = 17$

- A. $p = {}^-3,$ $q = 4$
- B. $p = {}^-4,$ $q = {}^-3$
- C. $p = 4,$ $q = {}^-3$
- D. $p = {}^-4,$ $q = 3$

356

1. Draw a pair of axes as shown, labelled *m* and *s*. The *s*-axis ranges from 0 to 10, while the *m*-axis ranges from 0 to 15. (Use a scale of 1 cm to 1 unit on both axes.)

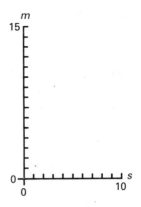

2. In Exercise 1, part A (p.353), the total height of a man and his son was 10 ft. This was written as $m + s = 10$.
   Plot the graph of $m + s = 10$.
   Use the graph to find:
   (a) *m* when $s = 2$,   (c) *s* when $m = 4$,
   (b) *m* when $s = 4.5$,   (d) *s* when $m = 7$.
   (*Note* All the points on the graph give all the possible solutions of the equation $m + s = 10$.)

3. In Exercise 1, part B (p.353), the total height of a man and two sons was 14 ft (the two sons were the same height). This was written as $m + 2s = 14$.
   Plot the graph of $m + 2s = 14$ using the same pair of axes as for question 2. Use the graph to find:
   (a) *m* when $s = 2$,   (c) *s* when $m = 4$,
   (b) *m* when $s = 4.5$,   (d) *s* when $m = 7$.
   (*Note* All the points on this graph give all the possible solutions of the equation $m + 2s = 14$.)

4. At the point of intersection of the two graphs (drawn in questions 2 and 3), what is the value of: (a) *s*?   (b) *m*?
   (*Note* The two equations have been solved simultaneously.)

5. Compare the answer to question 4 with the answer obtained in Exercise 1, parts C and D (p.354) and write what you notice.

Exercise 5

1. (a) List possible solutions for the equations:
   $x - y = 2$ and $2x + y = 10$
   Set them out as follows:

Solutions for $x - y = 2$		Solutions for $2x + y = 10$	
x	y	x	y
2	0	2	6
$2\frac{1}{2}$	$\frac{1}{2}$	$2\frac{1}{2}$	
3		3	
$3\frac{1}{2}$		$3\frac{1}{2}$	
4		4	
5		5	
6	4	6	
7		7	$^-4$
8		8	

(b) Compare the solutions for both equations and find the pair of solutions that is the same in both instances.

2. (a) Draw a pair of axes where the x-values range from 0 to 8 and the y-values range from $^-6$ to $^+6$. (Use a scale of 1 cm to 1 unit on both axes.)

   (b) Draw the graph of $x - y = 2$ (all the solutions of the equation $x - y = 2$, within the given range, lie on this line).

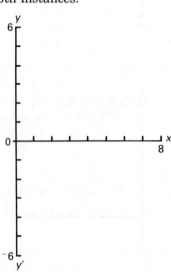

(c) Draw the graph of $2x + y = 10$ using the same pair of axes.
(All the solutions of the equation $2x + y = 10$, within the given range, lie on this line.)

(d) Find the point of intersection of the two graphs. Compare your answer with the answer to question 1(b), then write what you notice.

Exercise 6

Solve these pairs of simultaneous equations graphically. Draw a pair of axes as suggested by each sketch.

A

1.

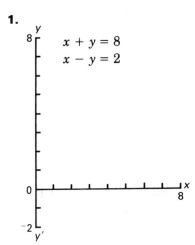

$x + y = 8$
$x - y = 2$

3.

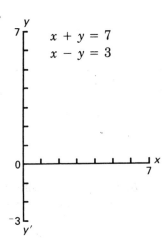

$x + y = 7$
$x - y = 3$

2.

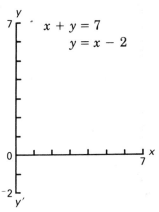

$x + y = 7$
$y = x - 2$

4.

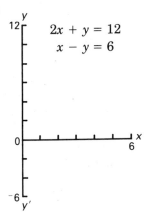

$2x + y = 12$
$x - y = 6$

5. $x + 2y = 10$
$x + y = 6$

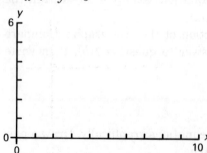

8. $x + 2y = 16$
$2x + y = 14$

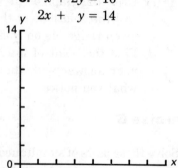

6. $4x + y = 8$
$x + y = 5$

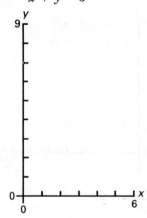

9. $2x + y = 8$
$2x - y = 2$

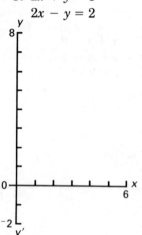

7. $2x + y = 18$
$x + y = 14$

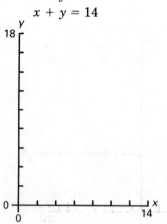

10. $x + 2y = 7$
$3x + 2y = 15$

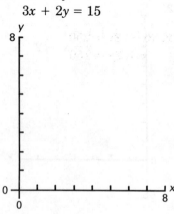

11. $x + 2y = 2$
$\quad\quad y = x + 8$

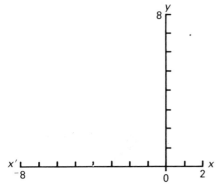

12. $x + y = {}^-6$
$\quad\quad x - y = 2$

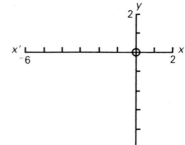

B **1.** $2x + y = 37$
$\quad\quad x + y = 24$

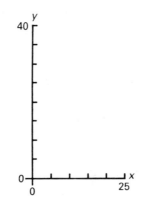

2. $3x + 2y = 290$
$\quad\quad x + 2y = 150$

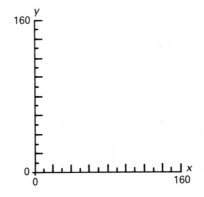

Exercise 7

For each problem, write two equations. Solve them simultaneously using the graphs.

1. (a) 2 pens and 1 watch cost £20. This can be written as $2p + w = 20$.
Draw the graph of $2p + w = 20$.

(b) 1 pen and 1 watch, of the same sort as those in part (a), cost £15. Form an equation.

(c) Draw the graph of the equation obtained in part (b).

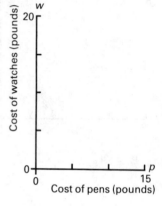

(d) What are the co-ordinates of the point of intersection of the two graphs?

(e) How much does one pen cost?

(f) How much does one watch cost?

2. A shop sold 2 sets of pliers and 3 sets of chisels at a total price of £90. On another day, they sold 4 sets of pliers and 3 sets of chisels at a total price of £120.

(a) Form two equations for the problem.

(b) Draw a pair of axes then draw graphs of the equations formed in part (a).

(c) Find the selling price of a set of pliers.

(d) Find the selling price of a set of chisels.

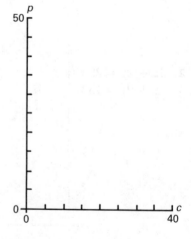

3. 3 containers of talcum powder and 6 bottles of splash-on cost £24, while 6 containers of talc and 2 bottles of splash-on cost £18. Find the cost of:

(*a*) one bottle of splash-on,

(*b*) one container of talc.

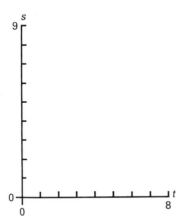

4. There are two types of pen on sale in a shop. 3 of type A and 5 of type B sell for £3.30, while one of each type sell for a total of 80 p. Find the selling price of each type of pen.

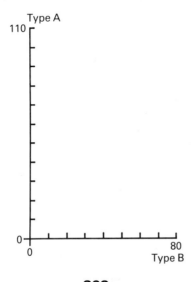

5. A shop sold two types of book: paperback and hardback. (All paperbacks cost the same and all hardbacks cost the same.) Find the cost of both types, if:

(a) 4 paperbacks and 3 hardbacks cost £30, while 2 paperbacks and 3 hardbacks cost £24.

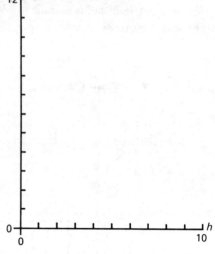

(b) 7 paperbacks and 2 hardbacks cost £35, while 7 paperbacks and 6 hardbacks cost £63.

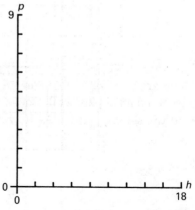

(c) 3 paperbacks and 5 hardbacks cost £48, while 4 paperbacks and 2 hardbacks cost £29.

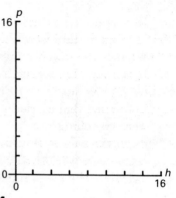

364

27 Statistics

Exercise 1   Bar Charts

1. The bar chart shows the number of people per car in cars travelling along a main road:

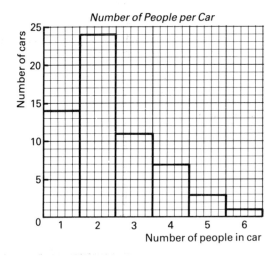

Number of People per Car

(a) How many cars carried 3 people?

(b) How many cars were there altogether?

(c) What is the modal number of people per car?

(d) In the cars that carried 2 people, how many people altogether were travelling?

(e) In the cars that carried 4 people, how many people altogether were travelling?

(f) In all the cars in the survey, what was the total number of people who were travelling?

(g) Calculate the mean number of people per car.

2. The following sums of money were donated to a charity by different people:

£6, £4, £9, £2, £7, £2, £3, £6, £2, £4,
£1, £5, £5, £3, £2, £1, £3, £5, £10, £6,
£2, £1, £4, £2, £5, £3, £5, £7, £6, £2

(*a*) Copy and complete the tally chart:

Amount	Frequency	Total
£1		
£2		
£3		
£4		
£5		
£6		
£7		
£8		
£9		
£10		

(*b*) Draw a bar chart to show the frequency distribution. Use a scale of 2 cm to £1 on the horizontal axis and 2 cm to 1 person on the vertical axis.

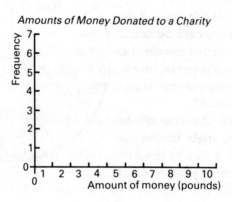

Amounts of Money Donated to a Charity

(c) How many people donated £5?

(d) What was the modal amount of money donated?

(e) How many people donated some money?

(f) What was the total amount of money donated?

(g) What was the mean amount of money donated?

(h) If two further donations of £8 are made, what is the new mean?

Exercise 2   Pictograms

Mr Blythe noted the number of cheerful people who spoke to him each day at work. He wrote the results in a table.

Day	Mon	Tue	Wed	Thur	Fri
Frequency	4	8	6	9	12

1. Copy and complete the table.

2. Draw a pictogram. Use ☺ to stand for 2 people.

3. Which day is probably 'pay day'?

Exercise 3   Pie Charts   M

1. The pie chart shows how 120 people travel to work.

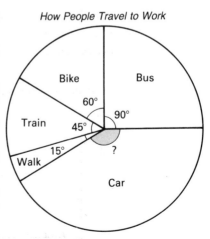

How People Travel to Work

(a) How many people travel by bike?

(b) What is the fraction of the 120 who travel by train? Write your answer as simply as possible.

(c) How many people travel on the train?

(d) What is the size of the missing angle in the car sector?

(e) How many people travel by car?

2. 2400 people were questioned in a survey.
To help in drawing a pie chart, a table was constructed as shown. The table was drawn up to relate the number of people with the angles needed at the centre of the pie chart.
(The first two entries show that 2400 people need an angle of 360° and 1200 people need 180°.)
Copy and complete the table.

Number of people	Angle	Number of people	Angle
2400	360°		270°
1200	180°	900	
600			36°
400			12°
	30°	160	
	15°	20	
	45°		21°
2000			330°
1000		180	
	120°		156°

3. In a survey of the type of holiday taken by 150 people, the following results were obtained:

Type of holiday	Number of people
Hotel	50
Guest house	25
Caravan	40
Camping	15
No holiday	?

(a) How many people did not have a holiday?

(b) Construct a pie chart to show the types of holidays taken.

Exercise 4   Jagged-Line Graphs

1. The graph shows a pupil's marks in 10 tests:

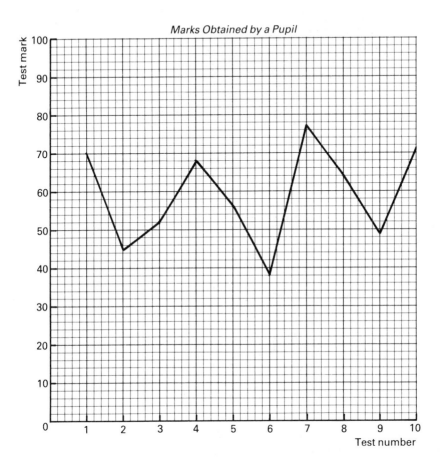

(a) Find the lowest test mark.

(b) Find the fourth highest test mark.

(c) Find the difference between the third highest and the second lowest marks.

(d) Find the mean test mark.

(e) Write the test marks in order from lowest to highest, then find the median.

(f) If each test mark was 8 marks higher, find both the new median and mean.

2. The table shows a firm's sales from the beginning of August to the end of March:

Month	Aug	Sept	Oct	Nov	Dec	Jan	Feb	Mar
Number sold (in 1000's)	30	35	43	46	49	65	74	82

(a) Draw a jagged-line graph to show the sales. Use a scale of 2 cm for each month and 2 cm for every 10 000 items sold.

(b) Between which two months was the greatest increase in sales?

(c) Explain how to recognise, on your graph, the greatest increase in sales.

(d) Calculate the mean number of items sold.

Exercise 5   Advantages and Disadvantages of Statistical Diagrams

Answer each question by selecting the most suitable diagram from—bar chart, jagged-line graph, pictogram or pie chart:

A *Advantages*

1. Which diagram is most likely to attract attention?

2. Which two diagrams are easiest to draw?

3. With which diagram is it easiest to compare the different component parts?

4. Which two diagrams are the most accurate to read?

5. In which diagram is it easiest to compare each part with the whole?

6. With which diagram is it easiest to show a large number of component parts without confusion?

7. Which diagram shows trend (that is, what might continue to happen)?

8. Which diagram is useful where actual values for each component part are not important?

B *Disadvantages*

1. For which diagram is a long time needed for it to be drawn well?

2. Which diagram is likely to be the most difficult to construct?

3. For which diagram are awkward calculations likely to be needed to be able to draw it?

4. In which two diagrams is it more difficult to accurately read the values of each component part?

5. Which diagram becomes much harder to read with a greater number of categories?

6. In which diagram is it difficult to compare component parts when there are only small differences between the component parts?

Misleading Presentations

Some diagrams printed in newspapers, in magazines and for advertising purposes can be both uninformative and misleading (sometimes deliberately so). It is necessary to study diagrams carefully and to pay particular attention to scales.

Exercise 6

The following four graphs show the same information. They all show the same increase in profit in one year.

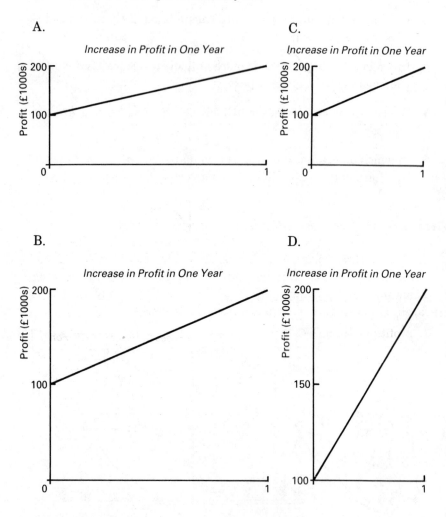

A.

Increase in Profit in One Year

C.

Increase in Profit in One Year

B.

Increase in Profit in One Year

D.

Increase in Profit in One Year

In all the graphs, last year's profit was £100 000 which has increased to £200 000 this year.

1. Which graph suggests the biggest improvement in business?

2. Which graph suggests the least improvement in business?

3. Explain how the different effects have been obtained.

Exercise 7

A The following diagrams are misleading in some way. Criticise each one.

1.

Sales

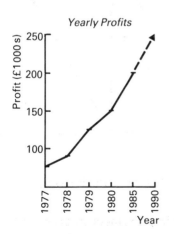

3.

Sales

2. Under new management, since 1980, look how our profits have soared:

Yearly Profits

4.

Houses Completed each Month

B Redraw the graphs in part A using correct scales.

373

Exercise 8

A firm advertised their interest rates with the following two graphs:

Borrow from
'Moneylend'.
Our interest rate is worth
consideration and you have
10 years to repay the loan.

Invest in
'Moneylend'
and make more.

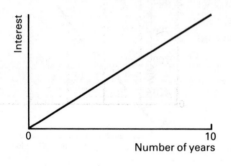
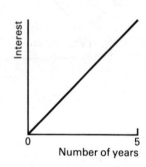

Criticise the above graphs explaining why they are misleading.

Exercise 9

1. (*a*) What is wrong with the following bar chart which shows a firm's profits over 8 years:

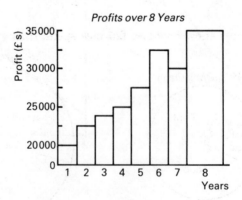

(*b*) Redraw the bar chart properly.

2. The graph shows the percentages of annual rainfall that fall in January in two places in Australia: Alice Springs and Sydney.

The graph suggests that it is much drier in Sydney!

(*a*) Why is the graph misleading?

(*b*) Calculate the January rainfall for each place if the annual rainfall for Alice Springs is 25 cm and for Sydney is 118 cm.

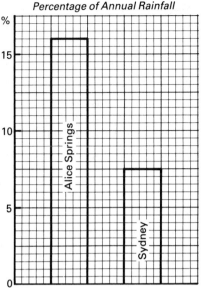

Percentage of Annual Rainfall

Exercise 10

Criticise the following statistical diagrams:

1. Record sales are getting bigger.

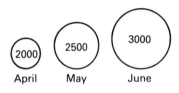

| April | May | June | July | August |

2. The pie charts show how two families spend their housekeeping money:

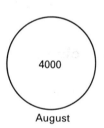

Mr and Mrs Clegg

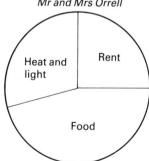

Mr and Mrs Orrell

Averages

Exercise 11

A Find the mode, median and mean of:

1. 2, 2, 8	**5.** 0, 0, 6	**9.** 3, 6, 9, 9, 12, 12, 19
2. 2, 8, 8	**6.** 0, 0, 0, 6	**10.** 5, 6, 8, 11, 12, 12
3. 6, 6	**7.** 2, 7, 13, 14, 14	**11.** 4, 6, 6, 8, 10, 14
4. 0, 6, 6	**8.** 4, 4, 8, 12	**12.** 4, 6, 6, 8, 10, 14, 400

B For each question, consider the statistical averages, mode, median or mean:

1. Which of the averages can have more than one value?

2. When the values are written in order of size, which of the averages has as many values before it as after it?

3. Which of the averages can be the greatest (or least) of the given set of values?

4. Which of the averages may not happen to be a number?

5. Which of the averages is affected by extreme values?

6. Which of the averages must appear in the given set of values?

Exercise 12

Each of the three lists of advantages and disadvantages given below describes one of the statistical averages: mean, median or mode. Decide which is which then write **1.** mean, and copy its list of advantages and disadvantages. Do likewise for **2.** median and **3.** mode.

A *Advantages*

   a It is unaffected by extreme values.

   b It can be found even if some of the values in the distribution may not be known.

Disadvantages

   a When a distribution is irregular this average may not appear to represent the set of values.

   b It cannot be used in further statistical calculations.

B *Advantages*

a Is an actual value in the data.

b It is unaffected by extreme values.

c It is easily obtained from a histogram.

Disadvantages

a There may be more than one of this average.

b When the data is grouped this average cannot be determined accurately.

c It does not necessarily appear to represent the set of values.

d It cannot be used in further statistical calculations.

C *Advantages*

a It can be calculated exactly.

b It makes use of all the data.

c It can be used in further statistical work.

d It is the best-known average.

Disadvantages

a It can have a silly value (e.g. 3.7 people).

b It is affected by extreme values.

c It cannot be obtained graphically.

Exercise 13

Which of the statistical averages, mode, median or mean, is most likely to have been used in the following?

1. The average shoe size.

2. The average distance travelled.

3. The batting average.

4. The average monthly rainfall.

5. The average size of jacket.

6. The average mark.

7. The average number of articles sold each month.

8. The average amount of money saved.

9. The average number of goals scored.

10. The average age.

Exercise 14

1. Nine pupils tried to estimate 1 min. Here are their estimates:
   67 s, 61 s, 54 s, 58 s, 53 s, 57 s, 63 s, 54 s and 62 s
   Find:
   (a) the mode,   (b) the median,   (c) the mean (to 1 d.p.).

2. In estimating the lengths of 8 lines, the errors were:
   8 cm, 2 cm, 6 cm, 2 cm, 9 cm, 7 cm, 6 cm, 2 cm
   Calculate the mean error.

3. It took me 7 throws at my first attempt to throw a six on an ordinary die. My next attempt only took 3 throws. Altogether, I threw 20 sixes. The number of throws needed each time were:
   7, 3, 4, 9, 4, 5, 7, 9, 6, 2, 10, 1, 5, 2, 5, 6, 11, 5, 1, 6
   Find:
   (a) the mode,   (b) the median.

4. Carry out an experiment of your own throwing a die to obtain twenty sixes. For each six thrown, note the number of throws needed. Find:
   (a) the mode,   (b) the median.

5. A cricketer scored 39, 87, 12, 0, 17, 103, 68 and 66 in eight innings. Calculate:
   (a) the mean score,
   (b) the mean score if the cricketer was not out in one of the eight innings.

   $$\left(\textit{Note} \text{ The average score} = \frac{\text{total score}}{\text{number of times out}}\right)$$

6. The attendance at seven football matches was:
   16 804, 12 392, 5418, 9265, 29 761, 25 424 and 14 910
   Find:
   (a) the median attendance,   (b) the mean attendance.

Exercise 15

1. Here are five numbers:
   6, 3, 8, 10, 3
   What additional number is needed to make the median of the six numbers equal to 5?

2. The mean mass of 7 people is 58 kg.

(*a*) Calculate their total mass.

(*b*) If 3 of these people have a mean mass of 54 kg, find:

    (i) the total mass of the other 4 people,

    (ii) the mean mass of the other 4 people.

3. The mean (average) age of 8 pupils is 15 years 5 months. Find their total age.

4. Frances got 32% and 54% in two tests. What percentage did she need in a third test for a mean (average) of 45%.

5. If 6 is the mean of 4, 5 and x, what is the value of x?

6. The mean of 2, 5, 9, n, n is 6. Find n.

7. Consider the numbers 2, 4, 7, 7, 7, 12, x, where x is bigger than 12. Where possible, find the:

(*a*) median,   (*b*) mean,   (*c*) mode.

8. Given the numbers 6, 9, 5 and 4, calculate:

(*a*) their sum,

(*b*) their mean,

(*c*) their mean if each number is increased by 3,

(*d*) their mean if each number is doubled.

9. A firm pays 2 people £164 per week and 3 people £189 per week. Calculate their mean weekly wage.

10. A firm pays 5 people £160 per week, 4 people £190 per week and 2 people £210 per week. Find:

(*a*) the modal wage,

(*b*) the median wage,

(*c*) the mean wage.

Range as a Measure of Dispersion

There are several ways of obtaining a measure of the spread (dispersion) of a set of values.

The *range* is one possible measure of dispersion. It is the difference between the highest and the lowest values. It is based entirely on the two most extreme values.

Exercise 16

1. Firm A pay a mean wage of £330 per week, while Firm B pay a mean wage of £210 per week. Which firm seems to be financially better to work for?

2. Two firms each have 5 employees. Firm A pays £130, £140, £180, £180 and £590 each week, while Firm B pays £190, £200, £220, £220 and £220 each week.
   (a) Which firm seems to be financially better to work for?
   (b) Calculate the mean weekly wage paid by each firm. Compare your answers with the figures in question 1 and comment on what you notice.
   (c) For each firm, find the range of the wages paid.

3. Two firms each have 7 employees. Firm X pays £150, £160, £170, £170, £220, £250 and £280 weekly, while Firm Y pays £50, £50, £60, £170, £170, £170 and £730 weekly.
   (a) Which firm seems to be financially better to work for?
   (b) Find the modal wage paid by each firm.
   (c) Find the median wage paid by each firm.
   (d) Find the mean wage paid by each firm.
   (e) For each firm, find the range of the wages paid.
   (f) Comment on your answers to this question.

Frequency Distributions

Exercise 17

1. The table shows the number of teams who scored the given number of goals on a certain Saturday:

Number of goals	0	1	2	3	4	5	6
Number of teams	40	44	32	7	5	2	1

   (a) Draw a histogram. Make each column 2 cm wide and use 2 cm to 5 teams on the vertical axis.
   (b) What is the mode?

2. The raw data below gives the length of time taken for each of 50 telephone calls. The times are given in minutes correct to the nearest minute.

```
12 14  7  2 29 44 15 13 18 16 21 13  8  6 10  4 17
17 24 32 38 11 14  2 27 21 34 11 13  6  7  2 16 37
37 15 16 24  1  7 14 13 22 28 13  8  7 30 12 23
```

(*a*) Construct a frequency distribution from the raw data given above. Set out the frequency table as follows.

Times	Tally	Frequency
0–5		
6–10		
11–15		
16–20		
21–25		
26–30		
31–35		
36–40		
41–45		

(*b*) Draw a histogram of this frequency distribution. Make each column 2 cm wide and use 1 cm to 1 unit for the frequency.

(*c*) Which is the modal class?

(*d*) What is the range of times?

In question 2, the *class intervals* used were 0–5, 6–10, 11–15, 16–20, 21–25, 26–30, 31–35, 36–40 and 41–45.

The end numbers for each class interval are called the *class limits*. The class limits for the first class, 0–5, are 0 and 5. 0 is called the *lower-class limit* and 5 the *upper-class limit*. For the fourth class, 16–20, 16 is the lower-class limit and 20 the upper-class limit.

Since all the times have been given to the nearest minute, the class interval 16–20 includes times that are less than 16 min as well as times longer than 20 min. The class interval 16–20 includes all the times between $15\frac{1}{2}$ min and $20\frac{1}{2}$ min. 15.5 (or $15\frac{1}{2}$) is called the *lower-class boundary* for that class interval while 20.5 ($20\frac{1}{2}$) is the *upper-class boundary*.

Exercise 18

1. Here are some scores at darts (from 3 throws):

> 120, 100, 66, 41, 121, 57, 43, 180, 65, 100, 125, 85, 101, 121, 105, 180, 135, 100, 39, 25, 10, 100, 80, 120, 62, 100, 123, 180, 61, 125, 100, 180, 6, 105, 100, 100, 95, 64, 43, 81, 133, 83, 40, 41, 82, 121, 130, 80, 95, 56

(a) Construct a frequency distribution using the class intervals: 1–20, 21–40, 41–60, 61–80, 81–100, 101–120, 121–140, 141–160 and 161–180.

(b) Draw a histogram of this frequency distribution. Make each column 2 cm wide and use 1 cm to 1 unit for the frequency column.

(c) Which is the modal class?

(d) What is the range of the scores?

Exercise 19

A 1. The class boundaries for the class intervals 1–6, 7–12, 13–18, 19–24 are 0.5, 6.5, 12.5, 18.5, 24.5, respectively. What are the class boundaries for 1–4, 5–8, 9–12, 13–16?

2. If the class boundaries are 0.5, 8.5, 16.5, 24.5, 32.5 and 40.5, what are the class intervals?

3. If the class boundaries are 82.5, 87.5, 92.5, 97.5, 102.5, 107.5 and 112.5, what are the class intervals?

4. The class boundaries for the class intervals 0–7, 8–15, 16–23, 24–31 are 0, 7.5, 15.5, 23.5, 31.5, respectively.

(a) What are the class boundaries for 0–3, 4–7, 8–11, 12–15?

(b) What are the class intervals if the class boundaries are 0, 4.5, 9.5, 14.5, 19.5, 24.5 and 29.5?

B The playing time for each side of several LP's are given below to the nearest minute:

> 21, 19, 27, 23, 15, 20, 19, 17, 14, 18, 39, 24, 24, 20, 20, 18, 19, 14, 15, 23, 21, 19, 16, 14, 20, 18, 14, 23, 19, 17, 19, 21, 20, 34, 25, 19, 20, 20, 17, 21, 21, 18, 14, 15, 20, 18, 17, 17, 17, 21, 19, 19, 21, 15, 21, 20, 26, 17, 15, 34, 39, 25, 15, 21, 14, 29, 24, 20, 21, 15, 23, 17, 18, 18, 22, 34, 16, 17, 20, 21

1. Draw a tally chart using class intervals of 5 min. (Use the class boundaries 13.5, 17.5, 21.5, 25.5, 29.5, 33.5, 37.5 and 42.5.)

2. Draw a histogram. Make each column 1 cm wide and use a scale of 2 cm to 5 units for the frequency.

3. Which is the modal class?

4. Into which class interval would a time of 28 min be put?

5. Into which class interval would a time of 21 min 34 s be put?

Exercise 20   Scatter Diagrams

1. In a competition, two judges each awarded marks out of 20 to fifteen competitors as shown in the table:

Competitor number	1	2	3	4	5	6	7	8	9	10	11	12	13	14	15
Judge A	15	12	7	20	10	17	12	10	14	19	16	8	18	15	17
Judge B	8	6	3	16	7	11	8	5	7	13	11	2	14	10	14

(a) Draw a scatter diagram to show the marks awarded but do not draw the line of best fit yet.

(b) Calculate the mean mark awarded by each judge.

(c) Plot the mean marks on the scatter diagram and circle that point.

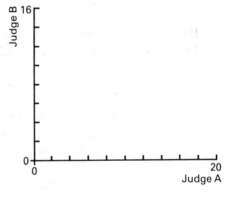

(d) Draw a line of best fit. (This line should pass through the point plotted from the two means.)

(e) Use the line to estimate the mark awarded by Judge B when Judge A awarded 12 marks.

(f) Use the line to estimate the mark awarded by Judge A when Judge B awarded 12 marks.

383

2. Answer the following questions using the scatter diagram below, which compares the times taken by 20 people to get to school with the distances they travel.

(*a*) How long is the journey likely to take for someone who travels 6 km?

(*b*) How long is the journey likely to take for someone who travels 10 km?

(*c*) A journey takes 10 min. What distance is likely to have been travelled?

(*d*) A journey takes 21 min. What distance is likely to have been travelled?

(*e*) How long is a journey likely to take for someone who travels 3.8 km?

(*f*) A journey takes 9 min. What distance is likely to have been travelled?

(*g*) A journey takes 27 min. What distance is likely to have been travelled?

(*h*) How long is a journey likely to take for someone who travels 2.7 km?

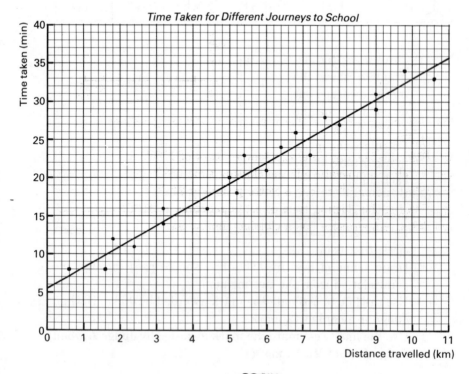

Time Taken for Different Journeys to School

3. Carry out a survey of distances travelled to school and total times taken by pupils in your class. Draw a scatter diagram.

4. The table below gives the clotting times of human blood plasma at various temperatures:

Temperature (°C)	20	32	10	15	25	18	28	16	8	18	12	22	19	12
Time (s)	50	17	75	65	42	52	24	53	82	55	76	52	60	62

Temperature (°C)	31	9	27	13	23	17	27	33	23	21	14	25	29
Time (s)	25	76	37	73	38	57	22	22	47	43	66	30	30

(a) Draw a pair of axes as shown and draw a scatter diagram. For this question, draw the line of best fit without first calculating and plotting the mean.

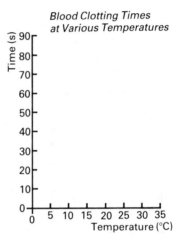

Blood Clotting Times at Various Temperatures

(b) Use the graph to find the clotting time at a temperature of 29 °C.

(c) Find the clotting time at a temperature of 12 °C.

(d) At what temperature is the clotting time 63 s?

(e) What temperature has a clotting time of 21 s?

(f) What is the clotting time at a temperature of 26 °C?

(g) What temperature has a clotting time of 9 s?

28 **Probability**

Exercise 1

1. A number is selected at random from the set of natural numbers from 1 to 50 inclusive. Find the probability that the number is:
   (a) even,
   (b) a square number,
   (c) a prime number,
   (d) a multiple of 3,
   (e) a multiple of 5,
   (f) a factor of 50,
   (g) greater than 30,
   (h) divisible by 1.

2. An unbiased coin is tossed four times. On the first three occasions it turns up heads. What is the probability that the fourth toss will also be heads?

3. An unbiased die has faces numbered 0, 1, 2, 4, 5, 5. What is the probability of obtaining:
   (a) a 5 in one throw of the die,
   (b) a number greater than 2 in one throw of the die,
   (c) a number less than 2 in one throw of the die.

4. A regular tetrahedron has the numbers 1, 2, 3 and 4 painted on its faces. Find the probability that when the tetrahedron is thrown like a die it lands with:

   (a) the 4 downwards,
   (b) an even number downwards,
   (c) a prime number downwards,
   (d) the sum of the three upper faces being even,
   (e) the sum of the three upper faces giving a prime number.

5. (*a*) In throwing an ordinary unbiased die, what is the probability of a 4?

(*b*) In throwing an ordinary unbiased die 150 times, how many times would you expect to get a 4?

6. There are 30 pegs in a box. $\frac{2}{5}$ are red, $\frac{1}{3}$ are green and the rest are blue.

(*a*) How many are blue?

(*b*) A peg is taken out of the box without looking, what is the probability that it is:

(i) red? (ii) green? (iii) blue? (iv) yellow?

7. There are 3 children. BGB stands for the first and third being boys and the second being a girl.

(*a*) Using the same three-letter notation, list all the possibilities for 3 children.

(*b*) If there are 3 children in a family, what are the chances of the children being all boys?

(*c*) If there are 3 children in a family, what is the probability that there are two girls and one boy?

(*d*) If there are 3 children in a family, what is the probability that there is at least one girl?

8. Out of 10 boys and 6 girls, what is the probability that one person selected at random will be a boy?

9. Out of 50 peaches, 4 were bad.

(*a*) What is the probability of picking a good peach at random?

(*b*) Out of a similar batch of 300 peaches, how many are likely to be bad?

(*c*) Out of a similar batch of 125 peaches, how many are likely to be bad?

10. There are 12 tins of the same shape and size and without labels. 8 of them are tinned pears, 3 contain tomato soup while the twelfth tin is a tin of beans. If I open one tin, what is the probability that for breakfast I shall have:

(*a*) pears?   (*c*) chicken soup?

(*b*) beans?   (*d*) tomato soup?

11. In the target shown, the probability of obtaining a certain score depends on the size of the angle at the centre.

What is the probability of scoring:

(a) 4?

(b) 6?

(c) 5?

(d) an odd number?

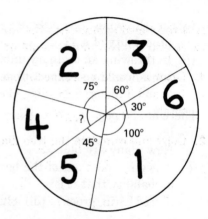

12. If the traffic lights show for the following times and a car approaches the lights, what is the probability that it will have to stop? The green light shows for 15 s, the amber for 3 s, the red for 25 s and the red and amber together for 2 s.

Exercise 2

A A four-edged blue spinner has the numbers 1, 3, 4 and 6 on it while a five-edged red spinner has the numbers 1, 3, 4, 4 and 7.

1. Copy and complete the table to show all the possible outcomes when both spinners are spun together and their scores totalled.

2. What is the probability of a total of 6?

3. What is the probability of a total of 4?

		Blue			
		1	3	4	6
Red	1				
	3				
	4		7		
	4				
	7			11	

4. What is the probability of a total of 3?

5. (a) What is the most likely total?

(b) What is the probability of the most likely total?

B A three-edged spinner has the numbers 1, 2 and 3 on it. It is spun twice and the two numbers obtained are totalled.

1. Draw a diagram similar to that in part A (but with the numbers 1, 2 and 3 along each side). Complete it to show all the possible totals.

2. Copy and complete the tree diagram:

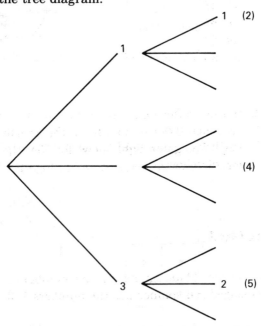

Totals

1   (2)

(4)

3   2   (5)

3. What is the probability of a total of:
   (a) 2?   (b) 3?   (c) 4?   (d) 5?   (e) 6?   (f) 7?

Exercise 3

A Copy the first five rows of Pascal's triangle, then complete the next three rows.

```
            1
          1   1
        1   2   1
      1   3   3   1
    1   4   6   4   1
  _   _   _   _   _   _
 _   _   _   _   _   _   _
_   _   _   _   _   _   _   _
```

B In tossing *1 coin* there is 1 way of getting 1 head and 1 way of getting 1 tail (0 heads). The probability of getting 1 head is $\frac{1}{2}$ and the probability of 0 heads $= \frac{1}{2}$.

These facts can be shown as follows:

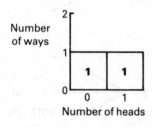

Number of heads	Probability
0	$\frac{1}{2}$
1	$\frac{1}{2}$

For *2 coins*, there is 1 way of getting 0 heads, 2 ways of getting 1 head (that is 1 head and 1 tail) and 1 way of getting 2 heads. This is shown as follows:

Number of heads	Probability
0	$\frac{1}{4}$
1	$\frac{2}{4} = \frac{1}{2}$
2	$\frac{1}{4}$

For *3 coins* we have:

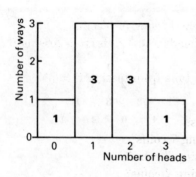

Number of heads	Probability
0	$\frac{1}{8}$
1	$\frac{3}{8}$
2	$\frac{3}{8}$
3	$\frac{1}{8}$

The probability of 2 heads and 1 tail $= \frac{3}{8}$.

For 4 coins we have:

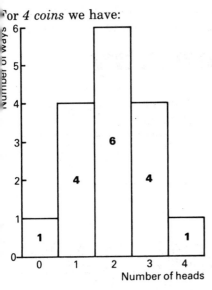

Number of heads	Probability
0	$\frac{1}{16}$
1	$\frac{4}{16} = \frac{1}{4}$
2	$\frac{6}{16} = \frac{3}{8}$
3	$\frac{4}{16} = \frac{1}{4}$
4	$\frac{1}{16}$

This table shows that the probability of 1 head and 3 tails $= \frac{1}{4}$.

1. From the graphs and tables, find the probability of obtaining:
   (a) 1 head and 2 tails from tossing 3 coins,
   (b) 3 heads from tossing 3 coins,
   (c) 2 heads and 2 tails from tossing 4 coins,
   (d) 3 heads and 1 tail from tossing 4 coins,
   (e) 4 tails from tossing 4 coins,
   (f) no tails from tossing 4 coins.

2. Compare the figures obtained in the graphs and tables with the numbers in each row of Pascal's triangle. Write what you notice.

3. (a) Draw a graph and table for the tossing of 5 coins.
   (b) What is the probability of 5 heads from tossing 5 coins?
   (c) What is the probability of 3 heads and 2 tails from tossing 5 coins?

4. (a) Draw a graph and table for the tossing of 6 coins.
   (b) What is the probability of 2 heads and 4 tails from tossing 6 coins?
   (c) What is the probability of no tails from tossing 6 coins?

5. What is the probability of:
   (a) no heads from tossing 7 coins?
   (b) 1 head and 6 tails from tossing 7 coins?
   (c) 2 heads and 5 tails from tossing 7 coins?
   (d) 3 heads and 4 tails from tossing 7 coins?

29 **Trigonometry**

Tangent

Exercise 1

A Find (correct to three decimal places):

1. tan 43°	**4.** tan 17°	**7.** tan 22.7°	**10.** tan 32.5°
2. tan 52°	**5.** tan 67°	**8.** tan 73.4°	**11.** tan 85.3°
3. tan 6°	**6.** tan 86°	**9.** tan 48.1°	**12.** tan 53.6°

B Find angle θ correct to one decimal place if:

1. tan θ = 0.4	**5.** tan θ = 1.047	**9.** tan θ = 0.0982
2. tan θ = 0.59	**6.** tan θ = 2.651	**10.** tan θ = 1.567
3. tan θ = 1.9	**7.** tan θ = 0.6128	**11.** tan θ = 2.1941
4. tan θ = 0.866	**8.** tan θ = 3.2449	**12.** tan θ = 42.8

Exercise 2

In each question, give angles correct to one decimal place and lengths correct to three significant figures:

1. $\triangle$ABC is right-angled at B.
AB = 8 cm and BC = 17 cm.
Calculate angle C.

2. $\triangle$STU is right-angled at T.
TU = 8 cm and angle
U = 37°.
Calculate the length of ST.

3. In △DEF, angle EDF = 90°
ED = 16 cm and DF = 9 cm.
Calculate angle E.

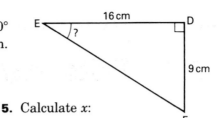

4. Calculate a:

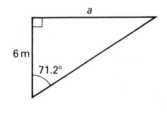

5. Calculate x:

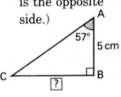

In right-angled triangle ABC, BC
is to be found. Either angle C or
angle A can be used.

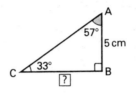

Using angle C
(The required
side, BC,
is the adjacent
side.)

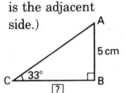

In right-angled △ABC,

since $\qquad \tan = \dfrac{\text{opp.}}{\text{adj.}}$

$$\tan \hat{C} = \frac{AB}{BC}$$

$$\tan 33° = \frac{5}{BC}$$

$$BC = \frac{5}{\tan 33°}$$

$$\underline{\underline{BC = 7.70 \text{ cm}}}$$

Using angle A
(The required
side, BC,
is the opposite
side.)

In right-angled △ABC,

since $\qquad \tan = \dfrac{\text{opp.}}{\text{adj.}}$

$$\tan \hat{A} = \frac{BC}{AB}$$

$$\tan 57° = \frac{BC}{5}$$

$$BC = 5 \tan 57°$$

$$\underline{\underline{BC = 7.70 \text{ cm}}}$$

Note It is easier to calculate the opposite side than the adjacent side.

Exercise 3

Calculate the required sides giving answers to three significant figures:

1.

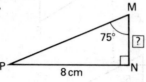

2.

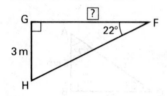

3.

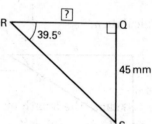

4.

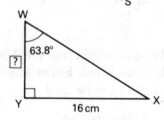

Exercise 4

1. CD = 9 cm and BC = 5 cm. Calculate the size of angle ADB.

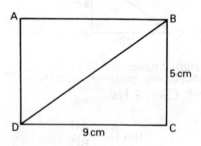

2. In the diagram, angle PRS = 141°, angle PQR = 51° and PR = 8.5 cm.

Calculate:
(*a*) angle PRQ,
(*b*) angle QPR,
(*c*) length PQ (correct to 3 s.f.).

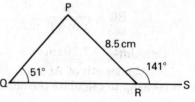

3. JKLM is a rectangle and P is a point on JK. JM = 10 cm, ML = 15 cm and ∠JMP = 46°.

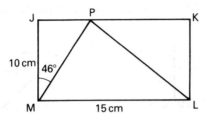

Calculate the length of:

(a) JP   (b) KP

4. Using the information given in the diagram:

(a) Calculate the length of RQ.

(b) Find SQ.

(c) Calculate angle PSQ.

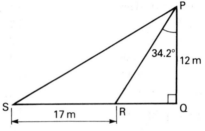

5. In the diagram, ∠ADC = 49°, ∠ACD = 90°, BC = 10 cm and CD = 8 cm.

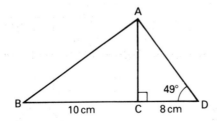

Calculate:

(a) the length of AC,

(b) angle ABC.

Sine and Cosine

Exercise 5

A Find, correct to four decimal places:

1. sin 65°
2. sin 71°
3. sin 28°
4. cos 25°
5. cos 87°
6. sin 59°
7. cos 43°

8. cos 56°
9. sin 13°
10. cos 8°
11. sin 33.5°
12. cos 78.8°
13. cos 8.7°
14. sin 82.9°

15. sin 47.2°
16. cos 19.6°
17. cos 62.4°
18. sin 69.7°
19. cos 36.1°
20. sin 21.3°

B Find angle θ correct to one decimal place if:

1. $\sin \theta = 0.9823$
2. $\sin \theta = 0.6198$
3. $\cos \theta = 0.5195$
4. $\sin \theta = 0.9239$
5. $\cos \theta = 0.2368$
6. $\cos \theta = 0.0941$
7. $\sin \theta = 0.3057$
8. $\cos \theta = 0.3616$
9. $\cos \theta = 0.6211$

10. $\sin \theta = 0.524$
11. $\sin \theta = 0.9977$
12. $\cos \theta = 0.7727$
13. $\cos \theta = 0.9681$
14. $\cos \theta = 0.849$
15. $\sin \theta = 0.7046$
16. $\sin \theta = 0.7815$
17. $\cos \theta = 0.6947$
18. $\sin \theta = 0.4524$

Exercise 6   **M**

1. Copy the table:

Angle θ	opp.	adj.	hyp.	$\dfrac{\text{opp.}}{\text{hyp.}}$	$\dfrac{\text{adj.}}{\text{hyp.}}$	$\cos \theta$	$\sin \theta$
10°							
30°							
40°							
45°							
70°							

2. For each triangle, measure the sides then complete the table:

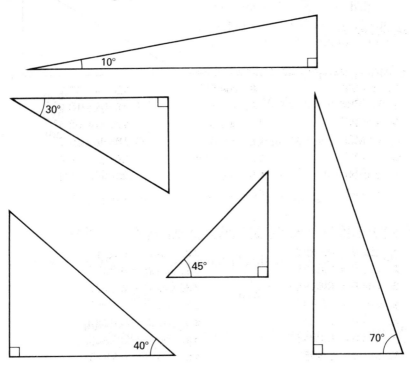

3. Compare the last four columns of your table. Write what you notice.

Exercise 7

A Draw three right-angled triangles of different sizes, where one angle in each triangle measures 35°. For each one, in relation to the 35° angle, measure and note the lengths of all three sides (adj., opp. and hyp.).

1. For each triangle, work out:

(a) $\dfrac{\text{opp.}}{\text{hyp.}}$ (b) $\dfrac{\text{adj.}}{\text{hyp.}}$

2. Using a calculator, find: (a) sin 35°, (b) cos 35°.

3. Write what you notice.

397

B **1.** Draw five different right-angled triangles that are not similar. Label each one LMN, where angle $M = 90°$.

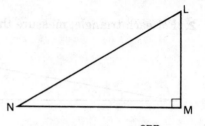

2. For each triangle:

(a) Measure angle N.

(b) Measure LM (opp.).

(c) Measure LN (hyp.).

(d) Measure NM (adj.).

(e) Work out $\dfrac{\text{opp.}}{\text{hyp.}}$

(f) Work out $\dfrac{\text{adj.}}{\text{hyp.}}$

(g) Find $\sin \hat{N}$.

(h) Find $\cos \hat{N}$.

3. Write what you notice about your answers to question 2.

From Exercise 7, you should have discovered that for any right-angled triangle, $\sin \theta = \dfrac{\text{opp.}}{\text{hyp.}}$ and $\cos \theta = \dfrac{\text{adj.}}{\text{hyp.}}$

We usually simply write: $\sin = \dfrac{\text{opp.}}{\text{hyp.}}$ $\cos = \dfrac{\text{adj.}}{\text{hyp.}}$

Exercise 8

A For each right-angled triangle, write, as a vulgar fraction, the value of the required ratios:

1. (a) $\sin \hat{A}$ (c) $\cos \hat{A}$
   (b) $\sin \hat{C}$ (d) $\cos \hat{C}$

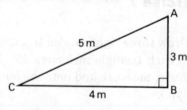

2. (a) $\sin \hat{L}$ (c) $\cos \hat{L}$
   (b) $\sin \hat{M}$ (d) $\cos \hat{M}$

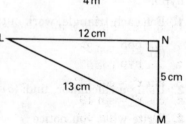

3. (*a*) $\sin \hat{Q}$ (*c*) $\sin \hat{R}$
(*b*) $\cos \hat{Q}$ (*d*) $\cos \hat{R}$

B For each right-angled triangle, work out the decimal values of the required ratios giving answers to four decimal places:

1. (*a*) $\sin \hat{K}$ (*c*) $\sin \hat{J}$
(*b*) $\cos \hat{K}$ (*d*) $\cos \hat{J}$

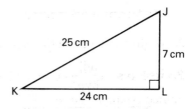

2. (*a*) $\sin \hat{S}$ (*c*) $\cos \hat{S}$
(*b*) $\sin \hat{U}$ (*d*) $\cos \hat{U}$

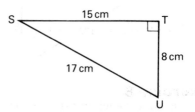

3. (*a*) $\sin \hat{X}$ (*c*) $\sin \hat{Z}$
(*b*) $\cos \hat{X}$ (*d*) $\cos \hat{Z}$

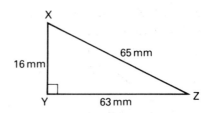

C Find the value of x if:

1. $\sin x° = \cos 38°$
2. $\sin x° = \cos 52°$
3. $\sin x° = \cos 74°$
4. $\cos x° = \sin 19°$
5. $\cos x° = \sin 45°$

6. $\sin 37° = \cos x°$
7. $\cos 63° = \sin x°$
8. $\cos x° = \cos 81°$
9. $\cos 26° = \sin x°$
10. $\cos x° = \sin 6°$

Exercise 9

In each right-angled triangle, find the required angle giving each answer in degrees (correct to one decimal place).

A **1.** Use $\sin = \dfrac{\text{opp.}}{\text{hyp.}}$

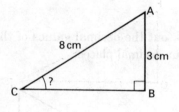

3. Use $\sin = \dfrac{\text{opp.}}{\text{hyp.}}$

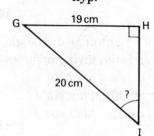

2. Use $\cos = \dfrac{\text{adj.}}{\text{hyp.}}$

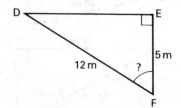

4. Use $\cos = \dfrac{\text{adj.}}{\text{hyp.}}$

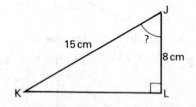

B Decide for each question whether to use $\sin = \dfrac{\text{opp.}}{\text{hyp.}}$ or $\cos = \dfrac{\text{adj.}}{\text{hyp.}}$

1.

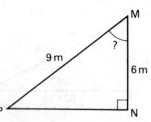

3.

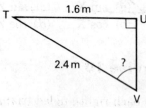

2.

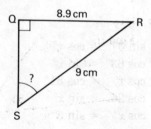

4.

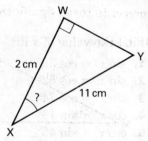

400

5.

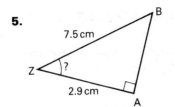

9.

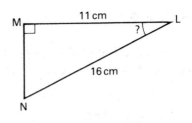

6.

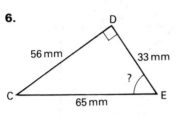

10.

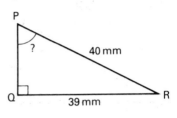

7.

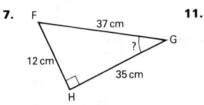

11.

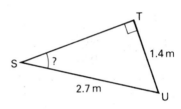

8.

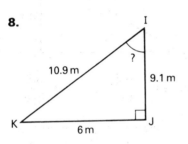

12.

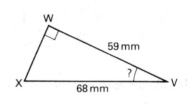

Exercise 10

In each right-angled triangle, find the required side giving answers correct to three significant figures.

A **1.** Use cos:

2. Use sin.
Find *d*.

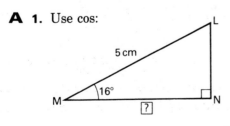

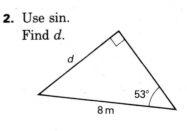

3. Use sin:

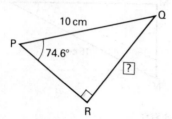

4. Use cos. Find x.

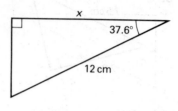

B Decide for yourself whether you must use sin or cos:

1. Find a:

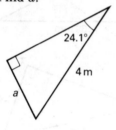

5. Find BC:

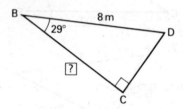

2. Find TU:

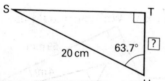

6. Find h:

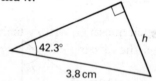

3. Find VX:

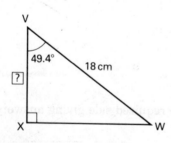

7. Find EF:

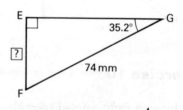

4. Find AY:

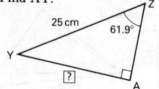

8. Find w:

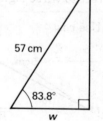

402

9. Find UT:

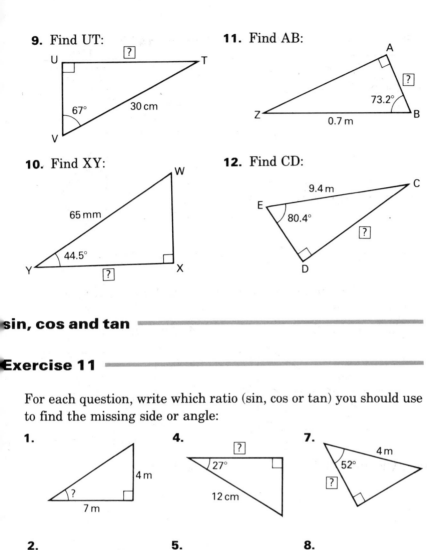

11. Find AB:

10. Find XY:

12. Find CD:

sin, cos and tan

Exercise 11

For each question, write which ratio (sin, cos or tan) you should use to find the missing side or angle:

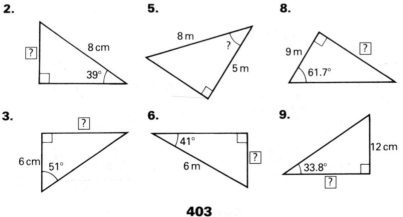

1.

4.

7.

2.

5.

8.

3.

6.

9.

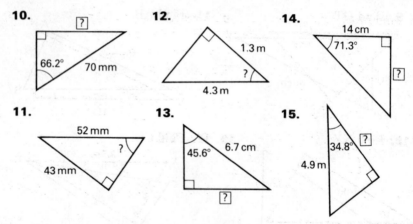

10. 66.2° 70 mm [?]

11. 52 mm 43 mm ?

12. 1.3 m 4.3 m ?

13. 45.6° 6.7 cm [?]

14. 14 cm 71.3° [?]

15. 34.8° 4.9 m [?]

Exercise 12

Calculate the required sides (to three significant figures) and the required angles (to one decimal place):

1. Find angle AIE:

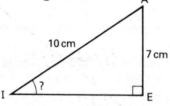

2. Find angle BJF:

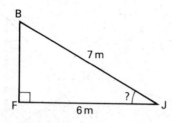

3. Find CG:

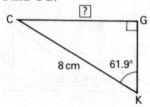

4. Find DL:

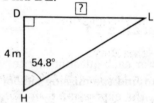

5. Find a:

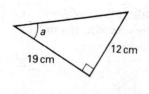

6. Find MU:

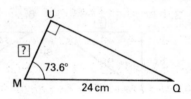

7. Find *b*:

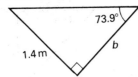

9. Find RV:

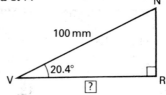

8. Find θ:

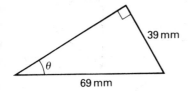

10. Find TX:

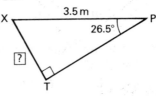

xercise 13

1. In the diagram, $\sin 35° = \dfrac{AB}{6}$

therefore, $\qquad AB = 6 \sin 35°$

so the length AB is given by the expression 6 sin 35°.
Write an expression for the length of side BC.

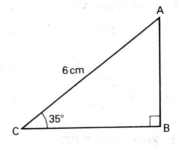

2. Write an expression for RQ involving sin, cos or tan.

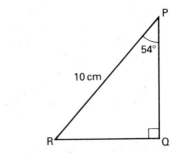

3. Which of the following is an expression for length XZ?

A. 7 sin 28°   D. $\dfrac{7}{\sin 28°}$

B. 7 cos 28°

C. 7 tan 28°   E. $\dfrac{7}{\tan 28°}$

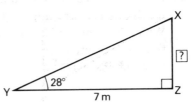

4. Select the correct two express-
ions for the length of MT:

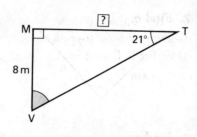

A. $8 \sin 21°$ E. $8 \tan 21°$

B. $8 \cos 69°$ F. $\dfrac{8}{\tan 21°}$

C. $8 \tan 69°$

D. $8 \sin 69°$ G. $\dfrac{8}{\sin 21°}$

5. Write an expression for the
length of:
(*a*) SR   (*b*) PR

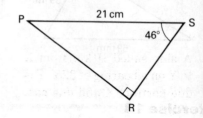

Exercise 14

In this exercise, give lengths correct to three significant figures,
bearings to the nearest degree and any other angles correct to one
decimal place:

1. A ladder, 6 m in length, is
placed against a vertical wall
and reaches 4.8 m up the
wall. What angle does the
ladder make with the
ground?

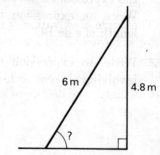

2. A ship sailed 32 km due
north then 41 km due east.
What is the bearing of
the end position from the
starting position?

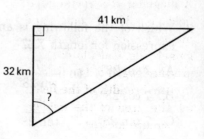

3. A ship sailed 30 km on a bearing of 062° from A to B. How far is B east of A?

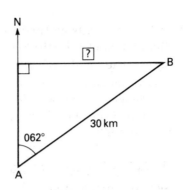

4. A ship sailed 10 km from X to Y on a bearing of 325°. Z is due north of X and due east of Y.
(a) How big is angle YXZ?
(b) Calculate distance XZ.

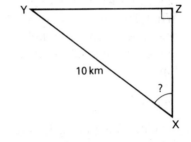

5. The angle of elevation of the top of a tower from a position on the ground 50 m from the foot of the tower is 26°. Calculate the height of the tower.

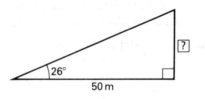

6. A diagonal of a rectangular field is 300 m long and makes an angle of 32° with a longer side. Calculate:
(a) the length of the field,
(b) the breadth of the field,
(c) the area of the field in square metres.

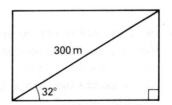

7. A boat at sea is 400 m from the bottom of a vertical cliff face. The cliff is 120 m high. Calculate the angle of depression of the boat from the top of the cliff.

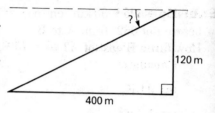

120 m

400 m

8. A rope is attached to both sides of a pair of step ladders, as shown, and is 1.1 m from the top of the ladder at the point where both sides touch. What length is the rope if the sides of the step ladder make an angle of 46° at the top?

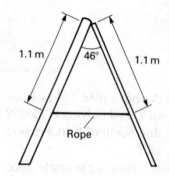

1.1 m   46°   1.1 m

Rope

9. A ladder, 7 m in length, rests against a vertical wall with its foot 3.5 m from the bottom of the wall. Calculate:
   (a) the angle the ladder makes with the ground,
   (b) the height the ladder reaches up the wall.

10. A trapezial table is shown:

89 cm

89 cm   89 cm

θ

181 cm

   (a) Calculate the angle marked θ.
   (b) Will two of these tables fit together to form a regular hexagon?
   (c) Calculate the perpendicular distance between the parallel sides.
   (d) Calculate the area of the table (in square metres).
      [Area of a trapezium = $\frac{1}{2}(a + b)h$]

408

1. In the diagram, $J\hat{L}M = 130°$, $J\hat{K}L = 40°$ and $LK = 5\,\text{cm}$.
Calculate:

(a) $J\hat{L}K$,

(b) $L\hat{J}K$,

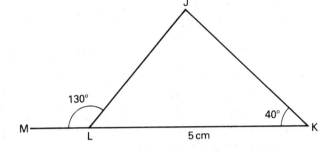

(c) the length of JK (giving your answer correct to three significant figures).

2. In $\triangle ABD$, C lies on BD such that $\angle ACB = 90°$. $AB = CD = 10\,\text{cm}$ and $\angle ABC = 64.1°$.
Calculate:
(a) the length of AC,
(b) $\tan \angle ADC$,
(c) the size of $\angle ADC$,
(d) the length of AD (use Pythagoras).

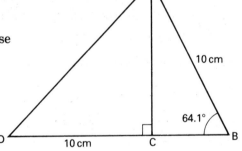

3. In the diagram, $EF = 12\,\text{cm}$, $E\hat{F}G = 35.7°$ and $H\hat{E}G = 44°$.
Calculate the length of:
(a) EG   (b) HG

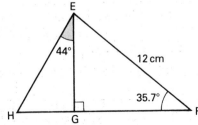

409

4. In the diagram, not drawn to scale, $\triangle$ WXZ is right-angled at W and WY is perpendicular to XZ where Y lies on XZ. WZ = 10 cm and WY = 7.5 cm.

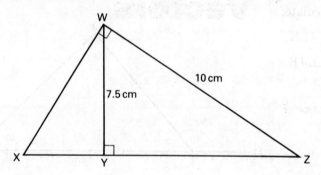

Calculate:
(a) angle WZX,   (b) the length of WX.

30 Vectors

Exercise 1

1. In the diagram, all the arrows show the vector $\begin{pmatrix} 4 \\ 2 \end{pmatrix}$.

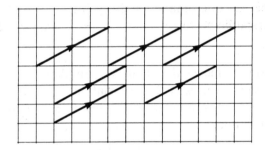

On squared paper, draw three arrows to show the vector:

(a) $\begin{pmatrix} 3 \\ 4 \end{pmatrix}$ (b) $\begin{pmatrix} ^-2 \\ 3 \end{pmatrix}$

2. In the diagram, A is the point (2, 4) and B the point (8, 1).
Which of the following is the vector $\overrightarrow{AB}$?

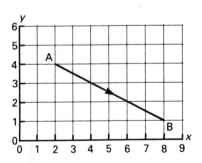

A. $\begin{pmatrix} ^-6 \\ 3 \end{pmatrix}$ B. $\begin{pmatrix} 3 \\ ^-6 \end{pmatrix}$ C. $\begin{pmatrix} 6 \\ ^-3 \end{pmatrix}$ D. $\begin{pmatrix} ^-3 \\ 6 \end{pmatrix}$

3. In the diagram, C is the point (8, 9) and D the point (3, 2).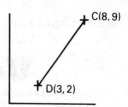
   (a) Which of the following is the vector $\overrightarrow{CD}$?

   A. $\begin{pmatrix} 5 \\ 7 \end{pmatrix}$ B. $\begin{pmatrix} 7 \\ 5 \end{pmatrix}$ C. $\begin{pmatrix} ^-5 \\ ^-7 \end{pmatrix}$ D. $\begin{pmatrix} ^-7 \\ ^-5 \end{pmatrix}$

   (b) Which of the vectors above is vector $\overrightarrow{DC}$?

4. A translation moves the point (2, 3) to (5, 7).
   (a) Write the vector of this translation.
   (b) Give the co-ordinates of the point (3, 0) under the same translation vector.

5. A translation moves the point (4, 1) to (2, 0).
   (a) Write the vector of this translation.
   (b) Give the co-ordinates of the point (6, 3) under the same translation.

6. A translation with vector $\begin{pmatrix} ^-2 \\ 3 \end{pmatrix}$ maps point E (4, 1) to the point F. What are the co-ordinates of F?

7. (a) Write the co-ordinates of G, H, I and J.
   (b) Vertex I is translated to the point (5, 2). What is the translation vector?
   (c) Write the images of G, H and I under the same translation.

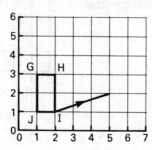

412

8. (a) Write the co-ordinates of K, L, M and N.

(b) Write the vector of the translation from trapezium P to trapezium Q.

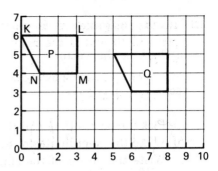

9. The vector $\begin{pmatrix} 4 \\ 6 \end{pmatrix}$ translates point P to Q. Which vector translates Q to P?

10. The vector $\begin{pmatrix} -5 \\ 3 \end{pmatrix}$ translates point R to S. Which vector translates S to R?

Exercise 2

1. Draw a pair of axes as shown. L is the point (0, 1). It is translated to M using the vector $\begin{pmatrix} 1 \\ 2 \end{pmatrix}$ and then from M to N using the vector $\begin{pmatrix} 2 \\ 5 \end{pmatrix}$.

(a) Plot the points L, M and N.

(b) Write the co-ordinates of N.

(c) Write the vector of the single translation that would move point L directly to N.

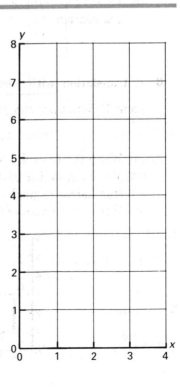

413

2. Draw a pair of axes as shown. P is the point (1, 0). It is translated

to Q using the vector $\begin{pmatrix} 4 \\ 1 \end{pmatrix}$ and then from Q to R using the

vector $\begin{pmatrix} -3 \\ 4 \end{pmatrix}$.

(a) Plot the points P, Q and R.
(b) Write the co-ordinates of R.
(c) Write the vector of the single translation that would move point P directly to R.

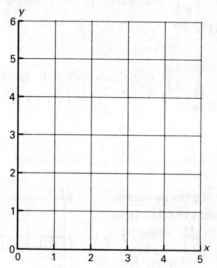

3. The vector $\begin{pmatrix} 5 \\ 4 \end{pmatrix}$ translates point S to T, while the vector $\begin{pmatrix} 4 \\ 3 \end{pmatrix}$ translates T to U.

(a) Which vector translates S directly to U?
(b) Which vector translates U directly to S?

4. The vector $\begin{pmatrix} 1 \\ -3 \end{pmatrix}$ translates point V to W while $\begin{pmatrix} 7 \\ 1 \end{pmatrix}$ translates W to X. Find the vectors:

(a) $\overrightarrow{VX}$ (b) $\overrightarrow{XW}$ (c) $\overrightarrow{WV}$ (d) $\overrightarrow{XV}$

5. Add these vectors:

(a) $\begin{pmatrix} 6 \\ 4 \end{pmatrix} + \begin{pmatrix} 3 \\ 2 \end{pmatrix}$ (b) $\begin{pmatrix} 7 \\ -1 \end{pmatrix} + \begin{pmatrix} 5 \\ 4 \end{pmatrix}$ (c) $\begin{pmatrix} 4 \\ -2 \end{pmatrix} + \begin{pmatrix} -2 \\ 6 \end{pmatrix}$

$(d) \begin{pmatrix} 5 \\ -6 \end{pmatrix} + \begin{pmatrix} -3 \\ 4 \end{pmatrix}$ $(g) \begin{pmatrix} 5 \\ -3 \end{pmatrix} + \begin{pmatrix} -5 \\ -4 \end{pmatrix}$ $(j) \begin{pmatrix} 9 \\ 2 \end{pmatrix} + \begin{pmatrix} -8 \\ 6 \end{pmatrix}$

$(e) \begin{pmatrix} -2 \\ 7 \end{pmatrix} + \begin{pmatrix} 4 \\ -4 \end{pmatrix}$ $(h) \begin{pmatrix} -2 \\ -7 \end{pmatrix} + \begin{pmatrix} -7 \\ -1 \end{pmatrix}$ $(k) \begin{pmatrix} 0 \\ -6 \end{pmatrix} + \begin{pmatrix} -6 \\ 6 \end{pmatrix}$

$(f) \begin{pmatrix} -6 \\ 3 \end{pmatrix} + \begin{pmatrix} 4 \\ -8 \end{pmatrix}$ $(i) \begin{pmatrix} 6 \\ 8 \end{pmatrix} + \begin{pmatrix} -9 \\ 0 \end{pmatrix}$ $(l) \begin{pmatrix} -5 \\ 2 \end{pmatrix} + \begin{pmatrix} 2 \\ -5 \end{pmatrix}$

Exercise 3

A Add these vectors:

1. $(a) \begin{pmatrix} 3 \\ 9 \end{pmatrix} + \begin{pmatrix} 5 \\ 2 \end{pmatrix}$ **4.** $(a) \begin{pmatrix} -1 \\ -5 \end{pmatrix} + \begin{pmatrix} 1 \\ 5 \end{pmatrix}$ **7.** $(a) \begin{pmatrix} -2 \\ -1 \end{pmatrix} + \begin{pmatrix} 2 \\ -1 \end{pmatrix}$

$(b) \begin{pmatrix} 5 \\ 2 \end{pmatrix} + \begin{pmatrix} 3 \\ 9 \end{pmatrix}$ $(b) \begin{pmatrix} 1 \\ 5 \end{pmatrix} + \begin{pmatrix} -1 \\ -5 \end{pmatrix}$ $(b) \begin{pmatrix} 2 \\ -1 \end{pmatrix} + \begin{pmatrix} -2 \\ -1 \end{pmatrix}$

2. $(a) \begin{pmatrix} 6 \\ 1 \end{pmatrix} + \begin{pmatrix} -2 \\ 3 \end{pmatrix}$ **5.** $(a) \begin{pmatrix} -3 \\ -4 \end{pmatrix} + \begin{pmatrix} 4 \\ -3 \end{pmatrix}$ **8.** $(a) \begin{pmatrix} -8 \\ 8 \end{pmatrix} + \begin{pmatrix} 8 \\ 8 \end{pmatrix}$

$(b) \begin{pmatrix} -2 \\ 3 \end{pmatrix} + \begin{pmatrix} 6 \\ 1 \end{pmatrix}$ $(b) \begin{pmatrix} 4 \\ -3 \end{pmatrix} + \begin{pmatrix} -3 \\ -4 \end{pmatrix}$ $(b) \begin{pmatrix} 8 \\ 8 \end{pmatrix} + \begin{pmatrix} -8 \\ 8 \end{pmatrix}$

3. $(a) \begin{pmatrix} -4 \\ 4 \end{pmatrix} + \begin{pmatrix} 3 \\ 6 \end{pmatrix}$ **6.** $(a) \begin{pmatrix} 8 \\ 0 \end{pmatrix} + \begin{pmatrix} -7 \\ -2 \end{pmatrix}$ **9.** $(a) \begin{pmatrix} 3 \\ -3 \end{pmatrix} + \begin{pmatrix} 2 \\ -2 \end{pmatrix}$

$(b) \begin{pmatrix} 3 \\ 6 \end{pmatrix} + \begin{pmatrix} -4 \\ 4 \end{pmatrix}$ $(b) \begin{pmatrix} -7 \\ -2 \end{pmatrix} + \begin{pmatrix} 8 \\ 0 \end{pmatrix}$ $(b) \begin{pmatrix} 2 \\ -2 \end{pmatrix} + \begin{pmatrix} 3 \\ -3 \end{pmatrix}$

B Add these vectors and show your results on a square grid:

1. $(a) \begin{pmatrix} 2 \\ 3 \end{pmatrix} + \begin{pmatrix} 4 \\ 1 \end{pmatrix}$ **2.** $(a) \begin{pmatrix} -2 \\ -3 \end{pmatrix} + \begin{pmatrix} -4 \\ -1 \end{pmatrix}$ **3.** $(a) \begin{pmatrix} 7 \\ 0 \end{pmatrix} + \begin{pmatrix} -3 \\ 6 \end{pmatrix}$

$(b) \begin{pmatrix} 4 \\ 1 \end{pmatrix} + \begin{pmatrix} 2 \\ 3 \end{pmatrix}$ $(b) \begin{pmatrix} -4 \\ -1 \end{pmatrix} + \begin{pmatrix} -2 \\ -3 \end{pmatrix}$ $(b) \begin{pmatrix} -3 \\ 6 \end{pmatrix} + \begin{pmatrix} 7 \\ 0 \end{pmatrix}$

C For each diagram, write the vector sum that is shown.

1. (a)

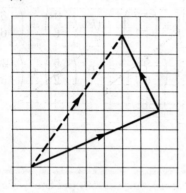

(b)

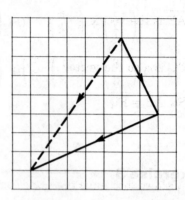

2. (a)

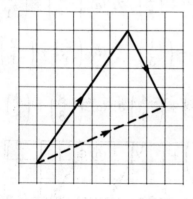

(b)

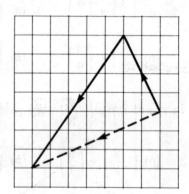

3. (a)

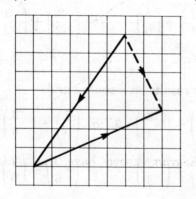

(b)

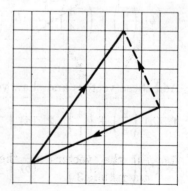

Equal Vectors

$\overrightarrow{AB}$ is the vector $\begin{pmatrix} 4 \\ -3 \end{pmatrix}$.

$\overrightarrow{PQ}$ is also the vector $\begin{pmatrix} 4 \\ -3 \end{pmatrix}$.

So $\overrightarrow{AB} = \overrightarrow{PQ}$.

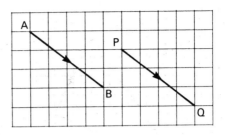

Vectors that have the *same length (same magnitude)* and the *same direction* are *equal.*

Exercise 4

In the diagram, $\overrightarrow{AB} = \overrightarrow{PQ}$. List any other vectors that are equal.

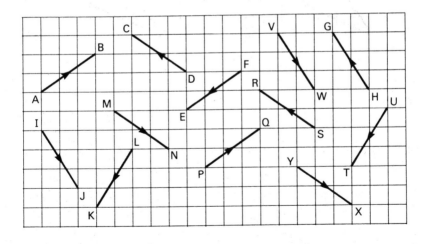

Note All the vectors in the diagram above have the same magnitude.

417

Further Addition of Vectors

Vector $\mathbf{b} = \begin{pmatrix} 3 \\ 1 \end{pmatrix}$.

In this diagram, all the arrows show vector $\mathbf{b}$.

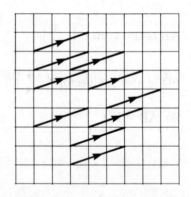

Vector $\mathbf{a} = \begin{pmatrix} 2 \\ 5 \end{pmatrix}$ and $\mathbf{b} = \begin{pmatrix} 3 \\ 1 \end{pmatrix}$.

In this diagram, only one arrow has been drawn for vector $\mathbf{b}$. The arrows could have been drawn anywhere. To add vectors $\mathbf{a}$ and $\mathbf{b}$, $\mathbf{a}$ can remain in the position shown but a different position for $\mathbf{b}$ must be used (or vice versa).

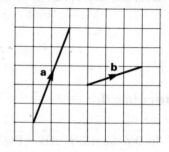

$$\mathbf{a} + \mathbf{b} = \begin{pmatrix} 2 \\ 5 \end{pmatrix} + \begin{pmatrix} 3 \\ 1 \end{pmatrix} = \begin{pmatrix} 5 \\ 6 \end{pmatrix}$$

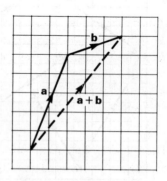

Note The position of $\mathbf{b}$ is chosen so that its arrow starts where the arrow for vector $\mathbf{a}$ ends.

Using the other notation, to find
$\overrightarrow{AB} + \overrightarrow{PQ}$:

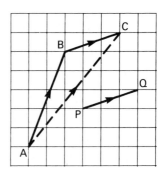

$$\overrightarrow{AB} + \overrightarrow{PQ}$$
$$= \overrightarrow{AB} + \overrightarrow{BC} \quad (\text{since } \overrightarrow{PQ} = \overrightarrow{BC})$$
$$= \overrightarrow{AC}$$

Note It is necessary to use $\overrightarrow{BC}$ instead of $\overrightarrow{PQ}$ since $\overrightarrow{AB}$ ends at B and $\overrightarrow{BC}$ starts at B.

Exercise 5

e.g. 1 Add the two vectors shown.
(Find $\overrightarrow{AB} + \overrightarrow{LM}$.)

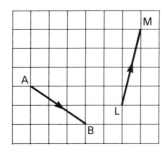 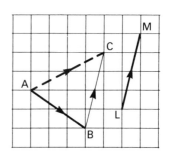

$$\overrightarrow{AB} + \overrightarrow{LM} = \overrightarrow{AB} + \overrightarrow{BC}$$

$$= \begin{pmatrix} 3 \\ -2 \end{pmatrix} + \begin{pmatrix} 1 \\ 4 \end{pmatrix}$$

$$= \begin{pmatrix} 4 \\ 2 \end{pmatrix}$$

419

e.g. 2 Find $\overrightarrow{PR} + \overrightarrow{PQ}$:

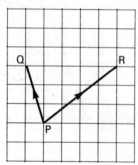

$$\overrightarrow{PR} + \overrightarrow{PQ} = \overrightarrow{PR} + \overrightarrow{RS}$$

$$= \begin{pmatrix} 4 \\ 3 \end{pmatrix} + \begin{pmatrix} ^{-}1 \\ 3 \end{pmatrix}$$

$$= \begin{pmatrix} 3 \\ 6 \end{pmatrix}$$

Add the given vectors:

1. Find $\overrightarrow{AB} + \overrightarrow{CD}$:

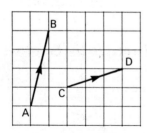

3. Find: (*a*) $\overrightarrow{CD} + \overrightarrow{EF}$
(*b*) $\overrightarrow{EF} + \overrightarrow{CD}$

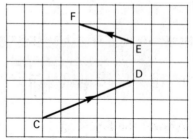

2. Find: (*a*) $\overrightarrow{EF} + \overrightarrow{GH}$
(*b*) $\overrightarrow{GH} + \overrightarrow{EF}$

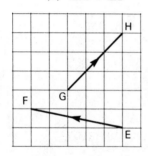

4. Find $\overrightarrow{CD} + \overrightarrow{FE}$:

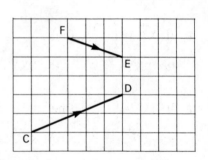

420

5. Find:
   (a) $\overrightarrow{GI} + \overrightarrow{GH}$
   (b) $\overrightarrow{GH} + \overrightarrow{GI}$

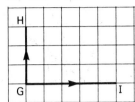

7. Find $\overrightarrow{NM} + \overrightarrow{NP}$:

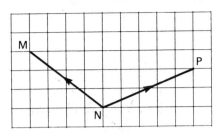

6. Find:
   (a) $\overrightarrow{JK} + \overrightarrow{JL}$
   (b) $\overrightarrow{JL} + \overrightarrow{JK}$

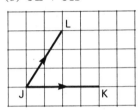

8. Find $\overrightarrow{QS} + \overrightarrow{QR}$:

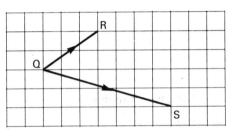

Exercise 6   Multiplication of Vectors by Scalars

A If $\overrightarrow{AB} = \begin{pmatrix} 3 \\ -4 \end{pmatrix}$ then $2\overrightarrow{AB} = \begin{pmatrix} 6 \\ -8 \end{pmatrix}$.

Answer the following:

1. If $\overrightarrow{PQ} = \begin{pmatrix} 2 \\ 3 \end{pmatrix}$, find:

   (a) $2\overrightarrow{PQ}$ (b) $5\overrightarrow{PQ}$ (c) $6\overrightarrow{PQ}$

2. If $\overrightarrow{XY} = \begin{pmatrix} 5 \\ -4 \end{pmatrix}$, find:

   (a) $2\overrightarrow{XY}$ (b) $3\overrightarrow{XY}$ (c) $7\overrightarrow{XY}$

3. If $\overrightarrow{TU} = \begin{pmatrix} -6 \\ 1 \end{pmatrix}$, find:

   (a) $3\overrightarrow{TU}$ (b) $4\overrightarrow{TU}$ (c) $10\overrightarrow{TU}$

4. If $\overrightarrow{VW} = \begin{pmatrix} ^-1 \\ ^-7 \end{pmatrix}$, find:

   (a) $4\overrightarrow{VW}$ (b) $5\overrightarrow{VW}$ (c) $6\overrightarrow{VW}$

5. If $\overrightarrow{KL} = \begin{pmatrix} ^-5 \\ 0 \end{pmatrix}$, find:

   (a) $2\overrightarrow{KL}$ (b) $3\overrightarrow{KL}$ (c) $8\overrightarrow{KL}$

B In the diagram, $\overrightarrow{WX}$ is equal to $2\overrightarrow{UV}$ but $\overrightarrow{ZY}$ is not equal to $2\overrightarrow{UV}$ ($\overrightarrow{ZY}$ points in a different direction):

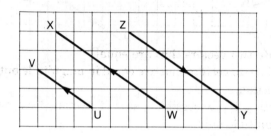

1. Which vector is equal to $\overrightarrow{AB}$?

2. Which vector is equal to $2\overrightarrow{AB}$?

3. Which vector equals $3\overrightarrow{AB}$?

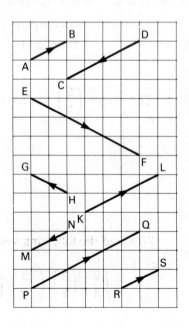

Exercise 7

1. The diagram shows several vectors:

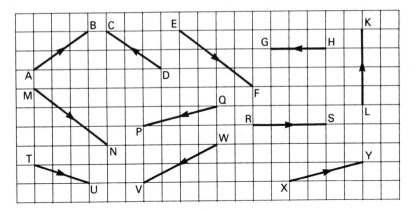

(a) Write two vectors that are equal.

(b) Write two vectors that are equal in magnitude but opposite in direction.

(c) Write all the possible pairs of vectors that are equal in magnitude only.

(d) Which is vector $\begin{pmatrix} 4 \\ 0 \end{pmatrix}$?

(e) Which vector equals $\overrightarrow{AB} + \overrightarrow{DC}$?

2. The vector $\begin{pmatrix} 3 \\ 2 \end{pmatrix}$ is shown.

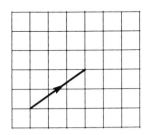

Which of the following vectors are perpendicular to vector $\begin{pmatrix} 3 \\ 2 \end{pmatrix}$?

$$\begin{pmatrix} 2 \\ 3 \end{pmatrix}, \begin{pmatrix} 3 \\ 2 \end{pmatrix}, \begin{pmatrix} 3 \\ -2 \end{pmatrix}, \begin{pmatrix} -2 \\ 3 \end{pmatrix}, \begin{pmatrix} -2 \\ -3 \end{pmatrix}, \begin{pmatrix} -3 \\ -2 \end{pmatrix}, \begin{pmatrix} 2 \\ -3 \end{pmatrix}, \begin{pmatrix} -3 \\ 2 \end{pmatrix}$$

3. The vector $\begin{pmatrix} -1 \\ 3 \end{pmatrix}$ translates

P to Q, while $\begin{pmatrix} 5 \\ 4 \end{pmatrix}$ translates

P to R.

Which vector translates:
(*a*) Q to R?   (*b*) R to Q?

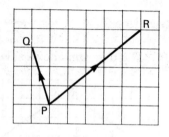

4. Vector $\begin{pmatrix} 6 \\ 2 \end{pmatrix}$ translates A to B, while $\begin{pmatrix} 4 \\ -3 \end{pmatrix}$ translates A to C.

Which vector translates:
(*a*) B to C?   (*b*) C to B?

5. Draw arrows on a square grid to show two translations that move a shape:
(*a*) equal distances but in opposite directions,
(*b*) equal distances in the same direction,
(*c*) different distances but in the same direction.

Revision Exercises XXIII to XXX

Revision Exercise XXIII

1. Using the diagram, find the bearing of:
  (a) Llandudno from Llangollen,
  (b) Llangollen from Llandudno.

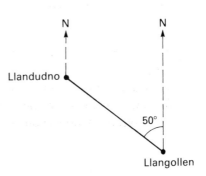

2. What is the bearing of:
  (a) P from O?   (b) Q from O?   (c) T from R?   (d) U from R?

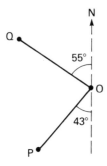
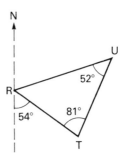

3. A ship is sailing on a bearing of 235°. Find its new bearing if it turns through 90° to starboard (the right).

4. A hiker walked on a bearing of 194°. What would be the bearing of the return journey?

5. In the diagram, the bearing of B from A is 062°, the bearing of C from A is 104° and $A\hat{C}B = 57°$.

Find:

(a) the size of $B\hat{A}C$,

(b) the bearing of A from C,

(c) the size of $A\hat{B}C$,

(d) the bearing of C from B.

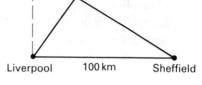

6. Sheffield is due east of Liverpool and 100 km away. Chorley is at a bearing of 040° from Liverpool and a bearing of 288° from Sheffield.

Make a scale drawing using a scale of 1 cm to 10 km.

Use your scale drawing to find:

(a) the distance of Chorley from Liverpool,

(b) the distance of Chorley from Sheffield,

(c) the bearing of Liverpool from Chorley,

(d) the bearing of Sheffield from Chorley.

7. A ship sailed from A to B, a distance of 40 km on a bearing of 225°. It then changed course to a bearing of 105° and sailed 80 km to C. Make a scale drawing. Use a scale of 1 cm to represent 10 km. Join AC.

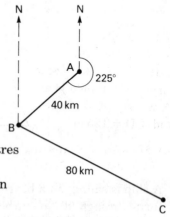

(a) Measure length AC in centimetres correct to one decimal place.

(b) Give the distance from C to A in kilometres.

(c) Measure $B\hat{A}C$.

(d) Give the bearing of C from A.

(e) Give the bearing of A from C.

(f) What is the size of $A\hat{C}B$?

8. Joe walked 6 km due east from K to L. He then walked to M on a bearing of 025° from L, a distance of 4 km. Make a scale drawing. (Use 1 cm to represent 1 km). By measuring your scale drawing, find:

(a) the direct distance from K to M,

(b) the bearing of M from K,

(c) the direction Joe needs to walk to return from M to K by the direct route,

(d) the total distance walked by Joe if he returned by the direct route,

(e) the time taken (to the nearest minute) to walk from K to L to M and then to return to K by the direct route at an average speed of 5 km/h.

9. From a point 40 m from the foot of a building, the angle of elevation of the top of the building is 15°. Make a scale drawing. Use the scale 1 cm represents 5 m.

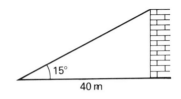

By measuring your scale drawing, find the height of the building.

10. The diagram, drawn to scale, shows two sides of a field:

AB = 80 m,
BC = 120 m and
A$\hat{\text{B}}$C = 70°. The fourth corner of the field is at D, where B$\hat{\text{C}}$D = 125° and CD = 125 m.

(a) Measure BC and use it to find the scale used. Copy and complete this sentence:
The scale is 1 cm represents ⎡ ? ⎤ m.

(b) Make a scale drawing of field ABCD. Use the same scale as that used in the diagram.

(c) Measure angle BAC.

(d) Give the length of AD correct to the nearest millimetre.

(e) Find the length of the side of the field, AD, in metres.

Revision Exercise XXIV

1. Which two rectangles are congruent?

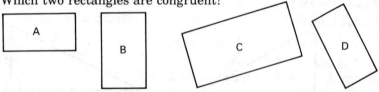

2. Write whether or not each given pair of triangles is congruent:

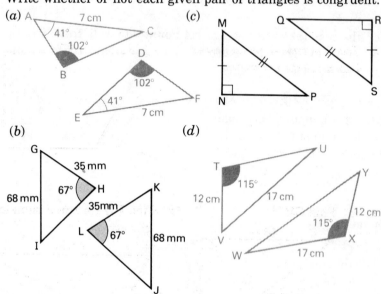

(a)

(c)

(b)

(d)

3. Name the two congruent triangles:

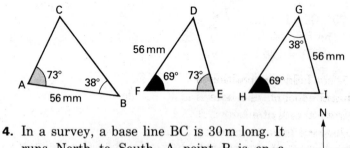

4. In a survey, a base line BC is 30 m long. It runs North to South. A point P is on a bearing of 115° from B and 060° from C. Make a scale drawing. Use the scale 1 cm represents 5 m. Find the distances BP and CP.

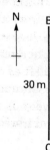

1. (*a*) Find CE:

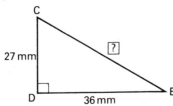

(*b*) Find TS (to 3 s.f.):

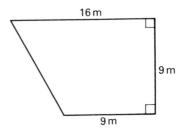

2. The diagram shows a garden. Calculate its perimeter.

3. (*a*) Find GH:

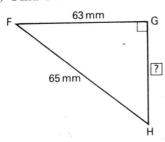

(*b*) Find PQ (to three significant figures):

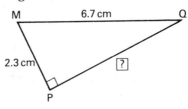

4. A 15-foot ladder is positioned so that the foot of the ladder is 5 ft from a wall of a house. If the distance from the ground to the bottom of the window is 14 ft and the distance from the ground to the top of the window is 18 ft, will the ladder hit the window?

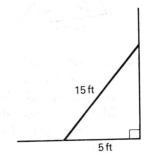

1. Which pair of values satisfies both equations simultaneously?

   (a) $k + 3h = 30$
   $k + h = 18$

   A. $k = 6, \ h = 8$
   B. $k = 9, \ h = 7$
   C. $k = 21, h = 3$
   D. $k = 12, h = 6$

   (b) $5q - 2r = 1$
   $5q + 3r = 11$

   A. $q = 3, \ r = 7$
   B. $q = 1, \ r = 2$
   C. $q = {}^-1, r = 2$
   D. $q = {}^-3, r = {}^-7$

2. Draw a pair of axes as shown. (Use a scale of 1 cm to 1 unit.)
   (a) Draw and label the lines which have the equations $x + y = 5$ and $y = x + 3$.
   (b) Using your graphs, solve the simultaneous equations:

   $x + y = 5$
   $y = x + 3$

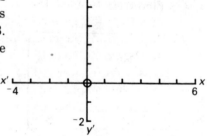

3. Draw a pair of axes as shown. (Use a scale of 1 cm to 5 units.)
   (a) A builder can build 4 houses and 2 bungalows in 80 weeks. This can be written as $4h + 2b = 80$. Draw the graph of $4h + 2b = 80$.

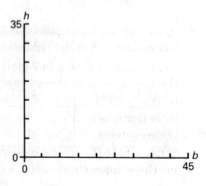

430

(*b*) The builder can build 2 houses and 3 bungalows in 60 weeks. Form an equation and draw its graph.

(*c*) Use the graphs to find the number of weeks taken to build a bungalow.

(*d*) Use the graphs to find the number of weeks taken to build a house.

Revision Exercise XXVII

1. Votes cast in an election were as follows:

Candidate	Number of votes
P. R. Ince	19 236
I. Askew	14 237
I. D. R. Down	4597

The electorate totalled 45 801.

(*a*) What was the total number of votes cast?

(*b*) How many of the electorate did not vote?

(*c*) What percentage of the electorate voted for I. Askew? (Give your answer to two significant figures.)

(*d*) Copy and complete the table showing the percentage of votes obtained and the angles needed to draw a pie chart:

Candidate	Number of votes	Percentage of electorate	Angle (to nearest degree)
P. R. Ince	19 236	42%	151°
I. Askew	14 237		
I. D. R. Down	4597		
Abstentions (non-voters)			

(*e*) Draw a pie chart using a circle of radius 5 cm.

431

2. The manager of a firm claimed that costs had reduced dramatically throughout the year and produced the graph below to prove the point.

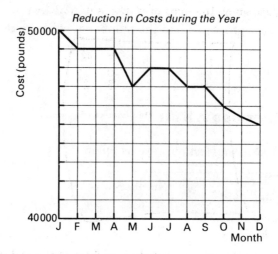

Reduction in Costs during the Year

(a) Explain why the graph is misleading.

(b) Redraw the graph as it should have been drawn.

3. Copy the following, but replace each question mark with one of the words: mode, median or mean:

(a) The ⬚?⬚ can have more than one value.

(b) The ⬚?⬚ is affected by extreme values.

4. The number of matches in ten boxes of matches was found to be:

43, 42, 39, 43, 41, 41, 40, 44, 43, 42

(a) What is the range of the number of matches?

(b) Which number is the mode?

(c) Find the median.

(d) Calculate the mean number of matches in a box. (Give the answer to the nearest whole number.)

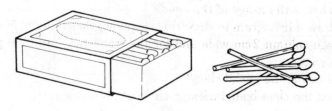

5. The mean (average) age of 4 people is 18 years. A fifth person, aged 23 years, joins the group. Find the new mean age.

6. The raw data show the ages of workers in a factory:

```
24  19   46   37  42
37  32   34   20  53
18  26   31   18  27
24  51   29   34  50
35  21   48   25  47
44  34   23   47
32  44   39   23
28  57   44   32
40  19   38   29
24  36   56   55
```

(*a*) Construct a frequency distribution from the raw data given above. Set out the frequency table as follows.

Ages	Tally	Frequency
16–20		
21–25		
26–30		
31–35		
36–40		
41–45		
46–50		
51–55		
56–60		

(*b*) What is the range of the ages?
(*c*) Draw a histogram to show this frequency distribution. Make each column 2 cm wide and for the frequency, use 2 cm per person.
(*d*) Which is the modal class?
(*e*) List the class boundaries.

433

7. The times taken (in minutes to the nearest $\frac{1}{10}$ min) for ten pupils to run two cross-country races are given in the table:

Pupil	1	2	3	4	5	6	7	8	9	10
First race	19.0	15.0	17.5	16.4	20.0	18.2	15.5	21.2	16.9	19.3
Second race	19.5	15.0	18.0	17.0	21.2	19.0	16.0	21.9	17.1	20.3

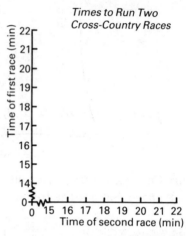

Times to Run Two Cross-Country Races

(a) Draw a pair of axes as shown. (Note the use of a broken scale.) Use a scale of 2 cm to 1 min on each axis.

(b) Plot the points given in the table to form a scatter diagram.

(c) Calculate the mean (average) time for each race.

(d) Plot the mean times on the scatter diagram and circle that point.

(e) Draw the line of best fit to pass through the point plotted from the two means.

(f) Use the line to estimate the time taken in the second race by someone who took 18.5 min in the first race.

(g) Use the line to estimate the time taken in the first race by someone who took 15.7 min in the second race.

Revision Exercise XXVIII

1. There are 16 girls' names and 20 boys' names in a bag. Find the probability that the first name drawn out of the bag is a boy's name.

2. A number is selected at random from the set of numbers {2, 3, 4, 8, 9, 12, 18, 25}. What is the probability that the selected number is a square number?

3. There are two, five-sided spinners. One has the numbers 1, 2, 2, 3, 5 on it, while the other has 1, 2, 4, 5, 6 on it.

Copy and complete the table to show all possible outcomes.

		First spinner				
		1	2	2	3	5
Second spinner	1		(2, 1)			
	2					
	4				(3, 4)	
	5					
	6	(1, 6)				

Use the table to find:
(a) the probability that the numbers obtained on spinning each spinner are the same,
(b) the probability of a total of 4,
(c) the most likely total,
(d) the probability of getting the most likely total,
(e) the probability of getting a total bigger than 7,
(f) the probability of getting a total that is an odd number,
(g) the probability that the number obtained on the first spinner is greater than that on the second,
(h) the probability that the number obtained on the second spinner is greater than that obtained on the first.

4. A four-edged spinner has the numbers 1, 2, 3 and 4 on it, while a three-edged spinner has 1, 2 and 3 on it.

Copy and complete the tree diagram showing all possible outcomes from spinning the two spinners.

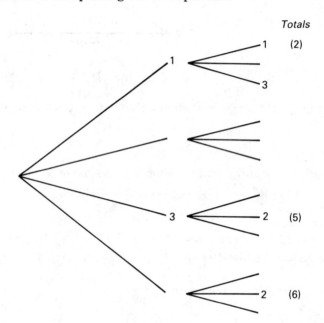

What is the probability of a total of:

(a) 2?   (b) 3?   (c) 4?   (d) 5?   (e) 6?   (f) 7?

5. One row of Pascal's triangle begins 1 5 10 and there are 6 numbers in that row. Write the full row.

Find the probability of obtaining:

(a) 5 tails from tossing 5 coins,

(b) 4 heads and 1 tail from tossing 5 coins.

Revision Exercise XXIX

1. Find, correct to three decimal places:

(a) $\tan 64.2°$ (b) $\sin 47.5°$ (c) $\cos 7.8°$

2. Find angle θ to one decimal place if:

(a) $\tan \theta = 1.46$ (b) $\sin \theta = 0.825$ (c) $\cos \theta = 0.565$

3. In the given triangle, which of the following is the value of tan BÂC?

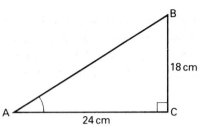

18 cm

A
24 cm
C

B

A. $\frac{3}{4}$ B. $\frac{4}{3}$ C. $\frac{3}{5}$ D. $\frac{5}{3}$ E. $\frac{4}{5}$

4. In right-angled triangle ACB shown above, which is the value of sin BÂC?

A. $\frac{3}{4}$ B. $\frac{4}{3}$ C. $\frac{3}{5}$ D. $\frac{5}{3}$ E. $\frac{4}{5}$

5. Find x if:
   (*a*) $\sin x° = \cos 40°$ (*b*) $\cos x° = \sin 69°$

6. In the diagram, CD = 5 cos 43°. Write an expression for the length of side:
   (*a*) BC   (*b*) AC   (*c*) CE

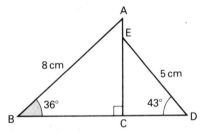

7. (*a*) Find AB (to three significant figures):

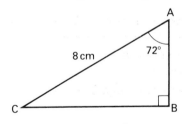

(*b*) Find ∠RPQ (to one decimal place):

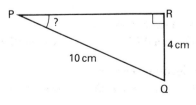

437

8. P is a point on the ground that is 45.3 m from the foot of a spruce. The angle of elevation of the top of the tree from P is 36°. Calculate the height of the tree (correct to one decimal place).

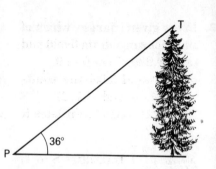

Revision Exercise XXX

1. Which of the following is vector $\overrightarrow{ED}$?

A. $\begin{pmatrix} 5 \\ 4 \end{pmatrix}$ C. $\begin{pmatrix} 5 \\ -4 \end{pmatrix}$

B. $\begin{pmatrix} 4 \\ 5 \end{pmatrix}$ D. $\begin{pmatrix} -4 \\ 5 \end{pmatrix}$

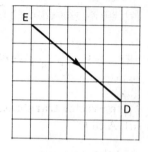

2. Triangle P is translated to position Q.
 (a) What is the translation vector?
 (b) What are the co-ordinates of the image of vertex C?

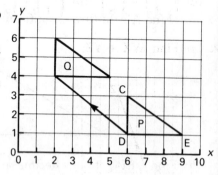

3. The vector $\begin{pmatrix} 3 \\ -7 \end{pmatrix}$ translates point A to B. Which vector translates B to A?

438

4. Draw a pair of axes where the
x-values range from 0 to 6 and
the y-values from 0 to 9.

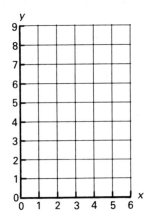

(a) Plot and label the points
R(4, 3) and S(1, 2).

(b) Which vector translates R
to S?

(c) If $\begin{pmatrix} 4 \\ 6 \end{pmatrix}$ translates S to T,
plot and label point T.

(d) What are the co-ordinates
of T?

(e) Which vector translates R
to T?

(f) Write the vector $\overrightarrow{TR}$.

5. The vector $\begin{pmatrix} -3 \\ 4 \end{pmatrix}$ translates A to B while $\begin{pmatrix} 7 \\ -5 \end{pmatrix}$ translates
B to C.

Which vector translates:

(a) A to C?   (b) C to A?

6. Add these vectors:

(a) $\begin{pmatrix} -6 \\ -2 \end{pmatrix} + \begin{pmatrix} 4 \\ 5 \end{pmatrix}$ (b) $\begin{pmatrix} 8 \\ -1 \end{pmatrix} + \begin{pmatrix} -2 \\ -2 \end{pmatrix}$ (c) $\begin{pmatrix} -4 \\ 4 \end{pmatrix} + \begin{pmatrix} -1 \\ -1 \end{pmatrix}$

7. (a) Find $\overrightarrow{WX} + \overrightarrow{YZ}$:   (b) Find $\overrightarrow{DE} + \overrightarrow{DF}$:

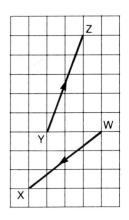

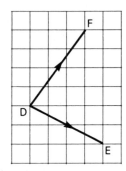

8. If $\overrightarrow{ST} = \begin{pmatrix} -2 \\ 4 \end{pmatrix}$ find:

    (a) $2\overrightarrow{ST}$ (b) $4\overrightarrow{ST}$ (c) $5\overrightarrow{ST}$

9. The diagram shows several vectors:

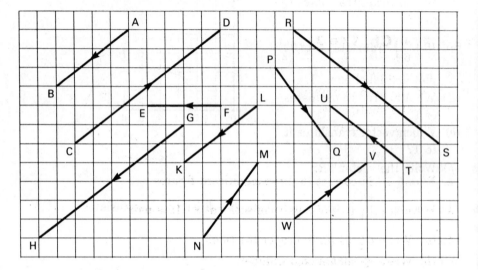

    (a) Write the names of two equal vectors.

    (b) Write the names of all the vectors that are equal in magnitude to $\overrightarrow{TU}$.

    (c) Which vector is perpendicular to $\overrightarrow{AB}$?

    (d) Which vector is perpendicular to $\overrightarrow{NM}$?

    (e) Which vector is equal to $2\overrightarrow{WV}$?

    (f) Which vector is equal to $2\overrightarrow{AB}$?

    (g) Write the name of a vector that is equal in magnitude to $\overrightarrow{AB}$ but opposite in direction.

Miscellaneous
Revision Papers 1–12

Paper 1 (Ch. 1 to 7)

1. From set P, list:
   (a) the prime numbers,
   (b) the multiples of 7,
   (c) the factors of 42.

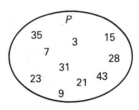

2. Round 4.185 correct to three significant figures.

3. Estimate the cost of 9 pairs of socks at £1.15 a pair.

4. (a) Write $6\frac{3}{4}$ as an improper fraction.
   (b) Write $\frac{35}{8}$ as a mixed number.

5. Two-thirds of a certain class are girls. If there are 18 girls, how many boys are there?

6. Change 4.83 km into metres.

7. A piece of cable, 15 ft long is cut from an 8-yard length. What length remains? (3 ft = 1 yd)

8. Calculate angle ADE:

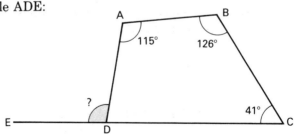

9. Use a calculator to find the following, correct to three significant figures:
   (a) 975^2 (b) 26.41^2 (c) 0.827^2 (d) 0.0418^2

10. Express in standard form ($a \times 10^n$ where $1 \leqslant a < 10$ and n is an integer):

(*a*) 3950   (*b*) 74 300   (*c*) 0.006   (*d*) 0.0149

Paper 2   (Ch. 1 to 7)

1. In a survey of 34 people, it was discovered that 15 bought biscuits yesterday, 26 bought biscuits or crisps or both, while 5 bought only biscuits and no crisps. Draw a Venn diagram and use it to find:

(*a*) the total number who bought crisps,

(*b*) the number who bought both crisps and biscuits,

(*c*) the number who bought neither crisps nor biscuits.

2. Which is the more accurate measurement:

(*a*) 8.45 m or 6.7 m?   (*b*) 3.4 kg or 5.12 kg?

3. Five-eighths of the pupils in the school watched TV last night. If there are 600 pupils in the school, how many did not watch TV?

4. Change 5820 mℓ into litres.

5. Which is the heaviest?

A. 0.04 kg   B. 4000 mg   C. 0.40 g   D. 0.4 kg   E. 40 g

6. Twelve 150 mℓ wine glasses are filled with wine. How many litres of wine are used?

7. Three angles of a quadrilateral measure 105°, 98° and 59°. Calculate the fourth angle.

8. Between which two consecutive whole numbers does $\sqrt{178}$ lie?

9. Give two numbers, each correct to one significant figure, between which these square roots lie:

(*a*) $\sqrt{87}$ (*b*) $\sqrt{800}$ (*c*) $\sqrt{7.69}$ (*d*) $\sqrt{5400}$ (*e*) $\sqrt{0.6}$

10. Use a calculator to find the following, correct to three significant figures:

(*a*) $\sqrt{51}$ (*b*) $\sqrt{87.6}$ (*c*) $\sqrt{9166}$ (*d*) $\sqrt{154.7}$ (*e*) $\sqrt{0.081}$

1. Multiply out:
  (a) $2(4x + 3)$ (b) $^-3(3c - 1)$

2. Factorise $16t + 4$.

3. Find the value of $12 \times 7.6 + 12 \times 2.4$ by factorising.

4. A net of a solid is shown:
  (a) What is the solid called?
  (b) Which point will meet point F when the solid is made?

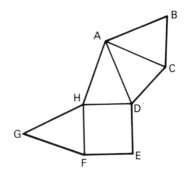

5. £154 is paid for 28 h work. What rate is that per hour?

6. Felicity earned £150 plus a bonus of 15% of that amount. Calculate the amount of bonus earned.

7. Aziz saves £4.25 each week. How much is that in one year?

8. A cuboid measures 12 cm by 8 cm by 4 cm.

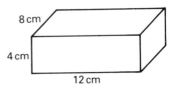

  (a) On squared paper, draw a net of the cuboid using a scale of 1 cm to 2 cm.
  (b) Calculate the area of the net you have drawn.
  (c) Calculate the total surface area of the full-size cuboid.
  (d) How many times as big as the area of the net is the area of the full-size cuboid?
  (e) Calculate the volume of the full-size cuboid.
  (f) Calculate the volume of the cuboid that would be made by the net you have drawn.
  (g) How many times as big as the volume of the cuboid made by the net is the volume of the full-size cuboid?

9. Draw a straight line PQ, 75 mm in length.

(a) Construct the locus of a point L, which moves so that LP = LQ.

(b) Draw the locus of a point M, which moves so that it is always 20 mm from PQ.

(c) Give the distances of the points of intersection of the two loci from P.

10. If the maximum loan a building society will allow is 80% of the cost of the house, find the maximum loan on a house costing £35 000.

11. A meal in a restaurant costs £18 plus VAT at 15%. Calculate:

(a) the amount of the VAT,

(b) the cost of the meal including VAT.

12. (a) Draw a pair of axes as shown.

(b) Plot the points A(5, 8) and B(5, 0) then join them with a broken line.

(c) Plot the points P(5, 7), Q(2, 4), R(3, 1) and S(5, 1), then join P to Q to R to S using straight lines.

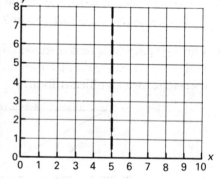

(d) Complete the figure so that it is symmetrical about AB.

(e) Write the name of the completed shape.

(f) Calculate the area of the completed shape.

Paper 4 (Ch. 8 to 15)

1. Simplify $5v - 2t - 2v + 6t - v$.

2. Factorise $36 + 9z$.

3. Simplify $5c - (3c + 2)$.

4. Here is the net of a polyhedron: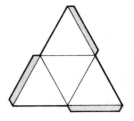
  (*a*) Name the polyhedron.
  (*b*) How many edges has it got?
  (*c*) How many vertices has it got?

5. A job pays a basic rate of £4.50 an hour. Calculate:
  (*a*) the basic wage for a 36-hour week,
  (*b*) the overtime rate at time and a half,
  (*c*) the overtime earned in 6 h at time and a half,
  (*d*) the total earnings from working a 42-hour week including 6 h overtime at time and a half.

6. How much per annum is £141 per week?

7. An article costs 8.60 kronor. What does it cost, in pence, if the exchange rate is 9.2 p to 1 kronor?

8. Part of a parallelogram is shaded. If $a = 3.18$ cm, $b = 6.91$ cm and $c = 2.65$ cm, calculate the shaded area giving your answer correct to two significant figures.

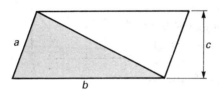

9. Regulations state that each person sleeping in a dormitory must have at least 15 m³ of air. Find the greatest number who are allowed to sleep in a dormitory measuring:
  (*a*) 10 m by 5 m by 3 m,
  (*b*) 12 m by 5 m by 3 m,
  (*c*) 12 m by 4 m by 2.5 m,
  (*d*) 12 m by 6 m by 3 m.

10. Mrs Shepherd earned £7800 a year. If she was given a 6% rise, calculate:
  (*a*) the amount of the increase,
  (*b*) Mrs Shepherd's new yearly earnings.

11. There were 480 people at a concert. If there were 120 seats not used, calculate the percentage attendance.

12. Find the simple interest on investing:
(a) £250 for 4 years at 8% p.a.,
(b) £294 for 6 years at $7\frac{1}{2}$% p.a.

13. A car bought for £4500 was sold at a loss of 20%. Calculate the selling price.

14. TP and TQ are tangents to a circle with centre O. If $Q\hat{O}T = 63°$, calculate $P\hat{T}Q$.

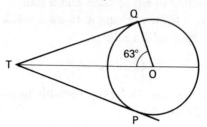

15. The sketch shows a pyramid fitted to the top of a cube.
(a) How many faces has the whole solid?
(b) How many edges has the whole solid?
(c) How many vertices has the whole solid?
(d) How many planes of symmetry has the whole solid?

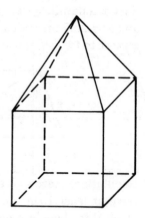

Paper 5 (Ch. 16 to 22)

1. If $y = 4x - 7$, find the value of y when:
(a) $x = 6$ (b) $x = 1$ (c) $x = 0$ (d) $x = {}^-5$ (e) $x = {}^-3$

2. If $y = x^2 - 12$, find the value of y when:
(a) $x = 2$ (b) $x = 10$ (c) $x = 0$ (d) $x = 1$ (e) $x = {}^-3$

3. $v = 30 - 10t$ gives the velocity $v\,\mathrm{m\,s^{-1}}$ of a stone, t s after it is thrown upwards at $30\,\mathrm{m\,s^{-1}}$. Use the formula to find:

(*a*) the velocity of the stone after $2\,\mathrm{s}$,

(*b*) its velocity after $1\frac{1}{2}\,\mathrm{s}$,

(*c*) the time it travels before it stops (it stops when its velocity, $v = 0$),

(*d*) its velocity after $5\,\mathrm{s}$ (explain the negative sign in this answer).

(*e*) Rewrite the formula to give t in terms of v.

(*f*) Copy and complete the following table for t and v:

Time, t s	0	1	2	3	4	5	6	7
Velocity, $v\,\mathrm{m\,s^{-1}}$		20						

(*g*) Draw a pair of axes as shown. t should range from 0 to 7. (Use a scale of 2 cm to 1 s.) v ranges from $^-40$ to 30. (Use a scale of 2 cm to $10\,\mathrm{m\,s^{-1}}$.) Draw a graph to show how v varies with t.

(*h*) Use the graph to find the time at which the stone is travelling downwards at $15\,\mathrm{m\,s^{-1}}$.

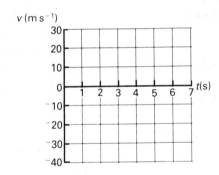

4. Rewrite the following formulae as instructed:

(*a*) $a + b + c = 90$, find c in terms of a and b.

(*b*) $A = 6\,lb$, find l in terms of A and b.

(*c*) $I = \dfrac{Prn}{100}$, find r in terms of I, P and n.

(*d*) $P = \dfrac{W}{A}$, find A in terms of P and W.

(*e*) $P = C + nd$, find n in terms of P, C and d.

5. Draw an enlargement of the given shape so that the perimeter is twice as long (the scale factor = 2).

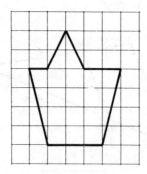

6. A rectangle measures 8 cm by 6 cm. Find in their simplest terms:
 (a) the ratio of the length to the breadth,
 (b) the ratio of the breadth to the perimeter,
 (c) the ratio of the perimeter to the length.

7. Mrs Ziegler earned £30 commission on sales totalling £500. At the same rate of commission, find:
 (a) the value of the sales when the commission was £150,
 (b) the value of the sales when the commission was £105,
 (c) the value of the sales when the commission was £78,
 (d) the commission on sales totalling £3500,
 (e) the commission on sales totalling £750,
 (f) the commission on sales totalling £1600.

8. Are the given pair of shapes similar or not?

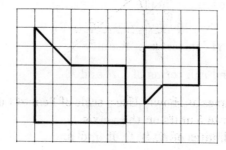

9. Triangles JKL and JNM are similar. JK = 4 cm, KN = 2 cm, JL = 6 cm and KL = 3 cm. Calculate:

(*a*) JM   (*b*) LM   (*c*) NM

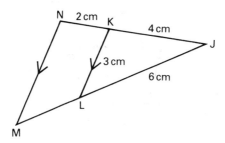

10. Solve the equations:

(*a*) $4x + 7 = 19$ (*b*) $2x - 3 = 4$ (*c*) $5 - 2x = 17$

11. If Alison got twice as many marks, she would have 18 more than Maurice who got 76. Let Alison's mark be x then form an equation in x. Solve the equation to find Alison's mark.

12. The graph of $x + y = k$ (where k is a constant) passes through the point (2, 7). What is the value of k?

13. The diagram shows the graphs of $x + y = 6$ and $y = 3x + 2$. One of them cuts the x-axis at A and the y-axis at B while the other cuts the x-axis at C and the y-axis at D. The graphs intersect at P.

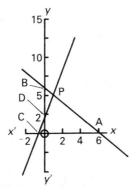

Draw a pair of axes as shown. Use a scale of 1 cm to 1 unit on the x-axis and 4 cm to 5 units on the y-axis.

Find the co-ordinates of A, B, C, D and P.

449

1. If $y = 5x - 2$, find the value of y when:
   (a) $x = 7$ (b) $x = 0$ (c) $x = 4$ (d) $x = {}^-2$ (e) $x = {}^-6$

2. If a turkey needs to be cooked for 15 min per pound plus an extra 15 min, for how long should a 9 lb turkey be cooked?

3. On a certain journey I travel at a steady 60 km/h.
   (a) If I travel s km in t h, copy and complete the following table showing s and t:

Time, t (h)	0	1	2	3	4	5	6	7
Distance, s (km)			120					

   (b) Using a scale of 2 cm to 1 h on the t-axis and 2 cm to 50 km on the s-axis, draw a graph to show how s varies with t.
   (c) Use the graph to find the time taken to travel 315 km.
   (d) Write a formula for s in terms of t.
   (e) Write a formula for t in terms of s.

4. Describe completely the transformation that maps:

   (a) A onto C   (d) C onto A
   (b) B onto C   (e) C onto B
   (c) A onto B   (f) B onto A

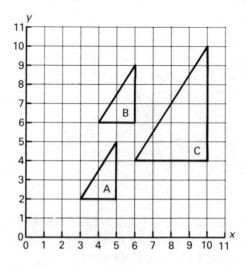

5. Draw an enlargement of the given shape so that each side is twice as long.

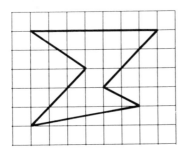

6. A model is 11.5 cm long. If the scale of the model is 1 to 72, find the true length.

7. At an average speed of 72 km/h a car completes its journey in 4 h.
  (*a*) How far is the journey?
  (*b*) At what average speed should the car have travelled to complete the journey in only 3 h?

8. Write whether or not each given pair of triangles is similar:
  (*a*)   (*b*)

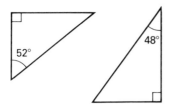 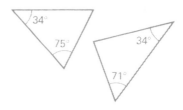

9. A church spire cast a shadow 68 m long, while a 2.1 m post cast a shadow 2.8 m in length. How tall is the church spire?

10. Solve the equations:
  (*a*) $5x - 2 = 28$
  (*b*) $19 - 2x = 10$
  (*c*) $3x + 14 = 2$

11. Give two values of x such that $3x > 21$ and $x < 9$.

12. Find the gradient of the line segment MN.

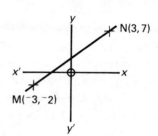

13. Six graphs are given below:

A.

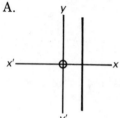

C.

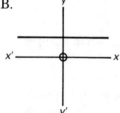

E.

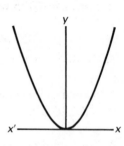

B.

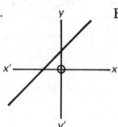

D.

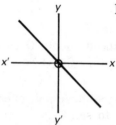

F.

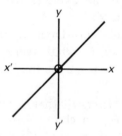

The equations of the graphs are:

$$y = {}^-2x, \quad y = 2, \quad y = 2x^2, \quad y = 2x, \quad y = x + 2, \quad x = 2$$

Which is the equation of:

(a) graph A?   (c) graph C?   (e) graph E?
(b) graph B?   (d) graph D?   (f) graph F?

Paper 7   (Ch. 23 to 30)

1. Using the diagram, find the bearing of:
   (a) Ballymena from Cookstown,
   (b) Cookstown from Ballymena.

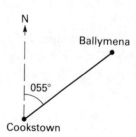

2. Make a scale drawing of field ABCD (use the scale 1 cm represents 30 m) given that AD = 90 m, ∠ADC = 95°, DC = 180 m, ∠BCD = 93° and BC = 165 m.

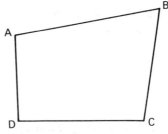

Find, by measuring,
(a) distance AB in metres,   (b) angle BAD.

3. Write whether or not the given pairs of triangles are congruent (equal angles are shown with the same markings):

(a)   (b)

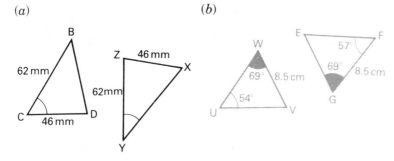

4. A ship sails 58 km on a bearing of 225° sailing from Port A to B. It then changes course to sail to C on a bearing of 135°, a distance of 51 km from B. Calculate the direct distance from A to C.

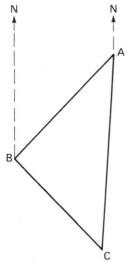

5. Draw a pair of axes as shown. (Use a scale of 1 cm to 1 unit.)

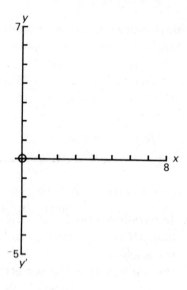

(*a*) Draw and label the line which has the equation $x + y = 7$.

(*b*) Draw and label the line which has the equation $x - y = 5$.

(*c*) Using your graphs, solve the simultaneous equations:
$$x + y = 7$$
$$x - y = 5$$

6. The table shows the number of mistakes per page made by a typesetter in typesetting 60 pages of a book:

Number of of errors	0	1	2	3	4	5	6	7	8
Frequency	3	7	12	16	7	5	6	2	2

(*a*) Draw a histogram. Make each column 2 cm wide. Use 1 cm to 1 unit for the frequency.

(*b*) What is the mode?

7. A card is drawn from a well-shuffled pack of 52 playing cards. What is the probability that:

(*a*) it is a number card bigger than 7?

(*b*) it is a red picture-card?

8. Find, correct to three decimal places:

(*a*) $\tan 33°$

(*b*) $\sin 27.3°$

(*c*) $\cos 68.1°$

9. In the diagram, which angle has a tangent of 0.5?

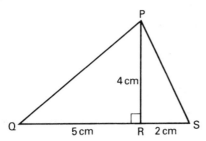

10. In parallelogram JKLM, JK = 8 cm, JM = 5 cm, ∠L = 125° and JN is perpendicular to LM. Find:
  (a) angle *M*,
  (b) the length of JN (correct to one decimal place),
  (c) the area of parallelogram JKLM (to the nearest square centimetre).

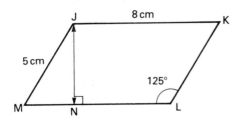

11. The vector $\begin{pmatrix} -1 \\ 5 \end{pmatrix}$ translates E to F. Which vector translates F to E?

12. The vector $\overrightarrow{GH}$ is shown. On squared paper, draw the vector $2\overrightarrow{GH}$.

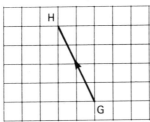

1. A hiker walked on a bearing of 293°. What would be the bearing of the return journey?

2. The direct distance between Shrewsbury and Stafford is 44 km, while the direct distance between Shrewsbury and Kidderminster is 49 km. Stafford is on a bearing of 076° from Shrewsbury, while the acute angle between the straight lines joining Stafford to Shrewsbury to Kidderminster is 60°.

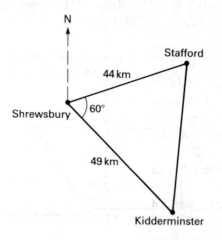

Make a scale drawing. Let 1 cm represent 10 km. Find:
(a) the bearing of Kidderminster from Shrewsbury,
(b) the bearing of Kidderminster from Stafford,
(c) the direct distance between Kidderminster and Stafford.

3. Name the two congruent triangles:

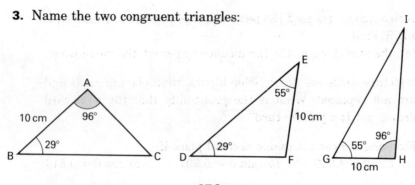

4. The plan is of a garden.

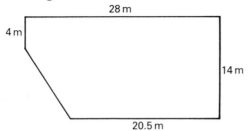

It would have been rectangular if it wasn't for a triangular piece missing. Calculate the perimeter of the garden.

5. Draw a pair of axes as shown. Use a scale of 1 cm to 1 unit on the c-axis and 1 cm to 2 units on the t-axis.

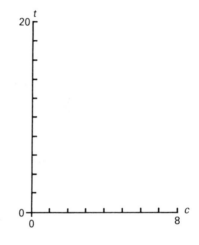

(a) A shopkeeper sold a toilet bag and 4 cosmetic cases (of the same cost) for £18. This can be written as $t + 4c = 18$. Draw the graph of $t + 4c = 18$.

(b) 3 toilet bags and 4 cosmetic cases (the same type as in part (a)) were sold for £30. Form an equation, then draw its graph.

(c) Use your graph to find the cost of a cosmetic case.

(d) Use your graph to find the cost of a toilet bag.

6. A firm pays 4 people £184 per week, 3 people £190 and 3 people £208. Find:
(a) the modal wage, (b) the median wage, (c) the mean wage.

7. 2 picture cards are drawn from a pack of 52 playing cards and are not replaced. What is the probability that the next card drawn out is a picture card?

8. Find angle θ correct to one decimal place if:
(a) $\tan \theta = 4.829$ (b) $\sin \theta = 0.997$ (c) $\cos \theta = 0.613$

9. △STU has a right-angle at T. SU = 25 cm and ∠SUT = 36.9°. Find:

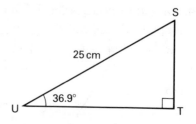

   (a) the length of ST (to the nearest centimetre),
   (b) the length of UT (to the nearest centimetre),
   (c) the area of triangle STU.

10. A vector translates the point P(4, 7) to P′(6, 3).
   (a) To which point will the same vector translate Q(⁻1, 5)?
   (b) If R′(5, 0) is the image of R under the same translation, what are the co-ordinates of R?

Paper 9   (Ch. 1 to 30)

1. In a survey, it was discovered that 39 people liked cabbage, 42 liked sprouts, 9 liked neither and 7 liked sprouts but not cabbage. Draw a Venn diagram and use it to find:
   (a) the number who liked only cabbage,
   (b) the number who liked both cabbage and sprouts,
   (c) the total number of people questioned in the survey if all of them gave replies.

2. The calculation 4.7×9.2 is estimated from 5×9. Work out the error.

3. Work out $5\frac{5}{6} \div 1\frac{3}{4}$ and simplify your answer.

4. After travelling 280 km, we had travelled three-sevenths of our journey. How far was the whole journey?

5. A circular piece of metal has a diameter of $2\frac{5}{16}$ in. Write this size as a decimal correct to three decimal places.

6. Find the two missing numbers in the following sequence:

   3,   ⁻6,   12,   ⁻24,   ⎡?⎤,   ⎡?⎤,   192

7. After cycling 6.4 km, how much further did Ray need to cycle to complete his journey of 10.2 km?

8. What is the volume, in cubic centimetres, of a bottle that holds 1.8 ℓ?

9. Give the readings on the following electricity meters in kW h. (Ignore the $\frac{1}{10}$ kW h digit.)

(*a*)

| 1 | 0 | 2 | 7 | 6 | | $\frac{1}{10}$ 1 | kW h |

(*b*)

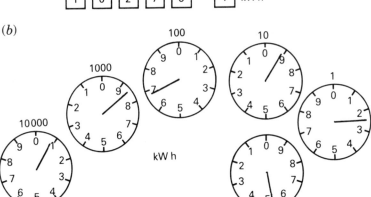

kW h

10. Draw a sketch of a cuboid.

11. The interior angles of a regular polygon measure 156°.
Calculate:

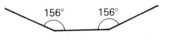

156°   156°

(*a*) each exterior angle,
(*b*) its number of sides.

12. Simplify $5x \times 3x$.

13. Write in standard form ($x \times 10^n$ where $1 \leqslant x < 10$ and n is an integer):
(*a*) 8820   (*b*) 813.7   (*c*) 0.702   (*d*) 0.09

14. Multiply out:
(*a*) $4(2t + 3)$ (*b*) $n(n + 3)$

15. Multiply out:
(*a*) $(c + 2)(c + 7)$ (*b*) $(x + 5)(x - 3)$

16. Factorise $8v - 28$.

17. Graeme earns £151.50 per week. How much does he earn in a year?

18. A cylindrical steel bar has a length of 1.2 m and a diameter of 3 cm.

    (*a*) What is the length of the bar in centimetres?

    (*b*) Calculate the volume of the bar in cubic centimetres. (Use $\pi = 3.142$.)

    (*c*) If the density of steel is 8000 kg/m³ (8 g/cm³), calculate the mass of the bar in kilograms. Give your answer correct to three significant figures.

19. A camera costs £175 plus 15% VAT. Calculate its cost inclusive of VAT.

20. TA and TB are tangents to a circle with centre O.

$\hat{\text{TBA}} = 72°$.

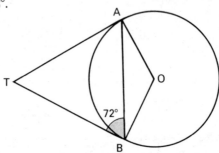

    (*a*) What sort of triangle is △TAB?

    (*b*) What sort of triangle is △AOB?

    (*c*) What sort of quadrilateral is TAOB?

    (*d*) Calculate angle BTA.

    (*e*) Calculate angle ABO.

    (*f*) Calculate angle AOB.

21.
```
X  X  X        X  X  X  X  X  X
X  •  X        X  •  •  •  •  X
X  X  X        X  X  X  X  X  X
```

One dot can be completely surrounded by 8 crosses as shown and 4 dots can be completely surrounded by 14 crosses.

    (*a*) Draw the pattern showing 3 dots completely surrounded by crosses.

    (*b*) How many crosses surround 2 dots?

    (*c*) How many crosses surround 5 dots?

    (*d*) How many crosses surround 20 dots?

(e) How many dots are surrounded by 20 crosses?

(f) Can dots be completely surrounded by 30 crosses?

(g) Can dots be completely surrounded by 25 crosses?

(h) How many dots are surrounded by 80 crosses?

(i) If there are d dots completely surrounded by x crosses, write a formula giving x in terms of d.

(j) Use the formula to find the number of crosses that completely surround 125 dots.

(k) Rewrite the formula giving d in terms of x.

(l) Use the re-arranged formula to find the number of dots that are completely surrounded by 100 crosses.

22. Using the diagram below, describe completely the transformation that maps:

(a) A onto B   (d) C onto E   (f) C onto F

(b) A onto C   (e) A onto D   (g) E onto F

(c) B onto E

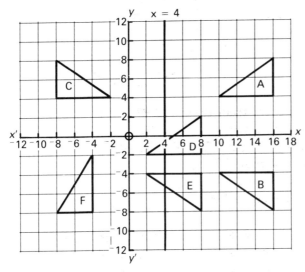

23. A model aeroplane is made to a scale of 1 to 72.

(a) If the wingspan of the model is 16.25 cm find the true size of the wingspan in metres.

(b) If the aircraft is 17.75 m long, find the length of the model in centimetres (to three significant figures).

24. Solve the equations:

(a) $3x + 5 = 26$ (b) $2x - 8 = 5$ (c) $6 - 5x = 21$

25. Bryan thought of a number, doubled it, then subtracted 9. The answer was 15. What was the number Bryan had thought of?

26. What is the gradient of line segment JK?

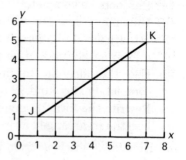

27. Using the diagram, find:
(a) the bearing of Blyth from Morpeth,
(b) the bearing of Morpeth from Blyth.

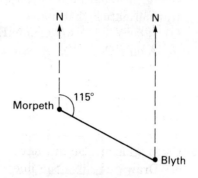

28. A.   C.   E.

B.   D.   F.

From the triangles, given above:
(a) Name two congruent triangles.
(b) Name two similar triangles that are not also congruent.

29. A boat is sailing across a river that is 56 m wide. Although it tries to sail straight across, the current sweeps it downstream so that the path followed is as shown.

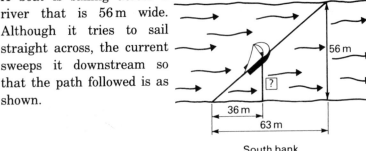

North bank

56 m

?

36 m

63 m

South bank

(a) How far is the boat from the south bank?
(b) How far will the boat have sailed in travelling from one bank to the other?
(c) How far has the boat sailed in reaching the position shown in the diagram?

30. Draw a pair of axes as shown.
($0 \leqslant x \leqslant 7$ and $^-2 \leqslant y \leqslant 6$.)
Use a scale of 1 cm to 1 unit.

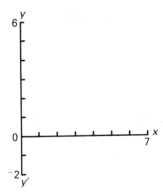

(a) Draw and label the line which has the equation $x + y = 6$.
(b) Draw and label the line which has the equation $y = x - 2$.
(c) Using your graphs, solve the simultaneous equations:
$$x + y = 6$$
$$y = x - 2$$

31. Bill Berry's mean (average) mark in two tests was 49%. Find his mean mark in three tests if in the third test he earned 64%.

32. A letter is chosen at random from the word DIFFERENCE.
(a) Which letter is most likely to be chosen?
(b) What is the probability that an F is chosen?
(c) What is the probability that a vowel is not chosen?

33. The diagram represents a vertical cliff CD which is 80 m high. A boat is sighted at B. Its angle of depression from D is 12°.

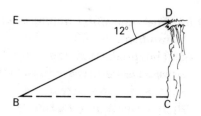

Calculate:

(a) angle BDC,

(b) the distance the boat is from the bottom of the cliff, giving the answer correct to the nearest metre.

34. The diagram shows several vectors:

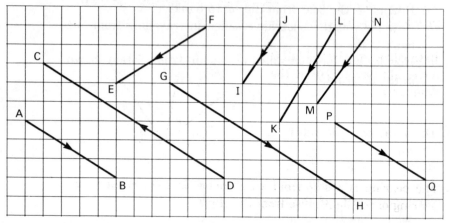

(a) Which two vectors are equal?

(b) Which vector is equal to $2\overrightarrow{AB}$?

(c) Which vector is perpendicular to $\overrightarrow{AB}$?

(d) Which vectors are equal in magnitude to $\overrightarrow{AB}$?

(e) Which vector is $\begin{pmatrix} -3 \\ -5 \end{pmatrix}$?

Miscellaneous Papers

Paper 10

1. For the set of numbers $\{1, 6, 9, 17, 21\}$ write whether each statement is true or false:

(a) There is exactly one prime number.

(b) One of the numbers has a factor of 4.

(c) The sum of the numbers equals the product of two of the numbers.

(d) The product of the second largest and second smallest numbers is greater than 100.

2. What is the square root of 81?

3. Calculate:
(a) $(4 + 3)(8 - 2)$ (b) $4 + 3(8 - 2)$

4. A bus set off at 10.51 and arrived at its destination at 14.07. How long did the journey take?

5. In the diagram given, PQR is a straight line. Calculate the value of x.

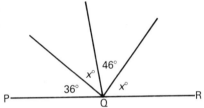

6. The bearing of a ship from a lighthouse is 230°. What is the bearing of the lighthouse from the ship?

7. PQ is a diameter of a circle with centre O. R lies on the circle and QP is produced to S. If $R\hat{P}S = 123°$, calculate $P\hat{Q}R$.

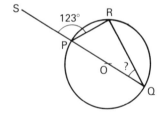

8. The probabilities that Beryl, Roy and Fiona will win a competition are $\frac{3}{8}$, $\frac{2}{5}$ and $\frac{9}{40}$ respectively. Who is most likely to win?

9. The petrol tank of Marina's car holds 54 ℓ when full. The car will travel 14 km per litre. If the tank is only two-thirds full, find:
(a) the number of litres of petrol in the tank,
(b) the total distance that can be travelled, if the tank is only two-thirds full at the beginning of a journey.

10. In the given diagram, QR is parallel to ST, $\widehat{PUT} = 75°$ and $\widehat{PTU} = 68°$. Find the size of:

(a) $\angle PVR$ (c) $\angle VPR$

(b) $\angle PVQ$ (d) $\angle VRT$

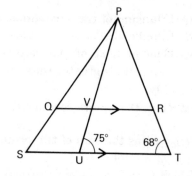

11. Use a calculator to find $\dfrac{3.9 \times 14.2}{1.8}$ correct to two significant figures.

12. (a) Follow the instructions in the given flow chart.

(b) Which of the following describes the list of numbers you have written?

A. triangular numbers

B. rectangular numbers

C. prime numbers

D. Fibonacci numbers

E. irrational numbers

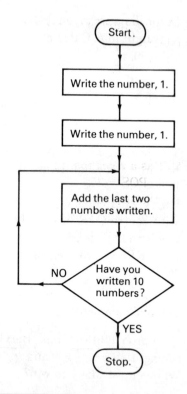

Paper 11

1. Write all the prime numbers between 1 and 15.

2. (a) How many millimetres are there in 1 cm?

(b) How many square millimetres are there in 1 cm²?

(c) Change 892 mm to centimetres.

(d) Change 7628 mm² to square centimetres.

(e) Change 2.41 cm² to square millimetres.

3. Simplify:

(a) $8x + 2x$ (b) $8x - 2x$ (c) $8x \times 2x$ (d) $8x \div 2x$

4. Write 14 as the sum of two prime numbers.

5. Which *two* of the following fractions are equivalent?

$$\frac{63}{81} \qquad \frac{30}{78} \qquad \frac{84}{147} \qquad \frac{84}{108}$$

6. Calculate the size of angle x in degrees.

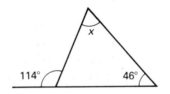

7. PQRS is a rhombus.
Angle PQS = 25°.
Give the sizes of:

(a) PÔQ   (c) PŜQ

(b) QŜR   (d) QP̂S

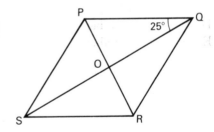

8. Copy and complete:

(a)

(b)

(c)

9. Consider the set of numbers $\{3^4, 6^3, 3^6, 9^2, 4^3, 2^9\}$.

    (*a*) Which member has the greatest value?

    (*b*) Which member has the least value?

    (*c*) Which two members have the same value?

10. Copy the following, but make each statement correct by inserting brackets:

    (*a*) $6 + 3 \times 2 = 18$ (*c*) $5 \times 8 - 7 + 2 = 31$

    (*b*) $4 \times 6 - 3 = 12$ (*d*) $5 \times 8 - 7 + 2 = 15$

11. A train journey lasts $3\frac{3}{4}$ h. Find the time of arrival if the train set off at 09.50.

12. In the diagram, O is the centre of the circle and $\angle RPQ = 52°$:

    (*a*) What is line PQ called?

    (*b*) What is line OR called?

    (*c*) What sort of triangle is $\triangle$ROP?

    (*d*) What is the size of angle ROP?

    (*e*) What is the size of angle PQR?

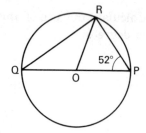

Paper 12

1. Change 2.49 m to centimetres.

2. Calculate:

    (*a*) 3.2×1.1 (*b*) $42.3 \div 9$ (*c*) $51.36 \div 20$

3. Find the sum of the first six positive odd numbers.

4. Estimate the number of millilitres of water in the jug.

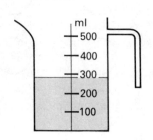

5. If * means 'multiply the first number by 3 then subtract the second number', find 7 * 9.

6. Which of the following statements are true for all parallelograms?
   A. Opposite sides are equal.
   B. The diagonals are equal.
   C. The diagonals bisect each other.
   D. Opposite angles are supplementary (i.e. they add up to 180°).
   E. The diagonals are perpendicular to each other.
   F. All the sides are parallel.

7. What is the size of angle *p* in the diagram?

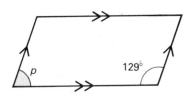

8. The graph of $3x + 4y = 24$ cuts the *x*-axis at P and the *y*-axis at Q. Find:
   (*a*) the co-ordinates of P,
   (*b*) the co-ordinates of Q,
   (*c*) the length of PQ.

9. (*a*) If 4 notepads cost the same as 3 ballpoint pens and 9 ballpoints cost the same as 4 packets of envelopes, how many notepads cost the same as 5 packets of envelopes?
   (*b*) If 5 notepads cost the same as 3 ballpoint pens, 5 ballpoint pens cost the same as 2 packets of envelopes and a notepad costs 18 p, find the cost of a packet of envelopes.

10. Natasha wanted to record 5 programmes on one 3-hour tape. Three of the programmes lasted 25 min each, one programme lasted 1 h 10 min and the fifth programme lasted half an hour. Did the programmes fit on the tape?—If so how many minutes of recording time were left?—If not, by how many minutes did the programmes overrun?

11. (*a*) How many strips, each 7.8 cm long, can be cut from a strip 1.5 m long?
   (*b*) If the strips were cut as in part (*a*), what length will be left over?

12. Mr Kendall travelled 50 km by car.
  (a) If he averaged 75 km/h for the first 25 km, how long did that part of the journey take him?
  (b) If he averaged 50 km/h for the remaining 25 km, how long did that part of the journey take him?
  (c) What was his total time for the whole journey?
  (d) What was his average speed for the whole journey?

Appendix 1
Irrational Numbers

A proof that $\sqrt{2}$ is an irrational number (p.75)

Assume $\sqrt{2}$ is a rational number.

We can therefore let $\dfrac{p}{q} = \sqrt{2}$ (where $\dfrac{p}{q}$ is in its simplest form, i.e. p and q have no common factor)

so $\left(\dfrac{p}{q}\right)^2 = 2$

or $\dfrac{p^2}{q^2} = 2$

Hence $p^2 = 2q^2$

Since the right-hand side is exactly divisible by 2 the left-hand side must also be exactly divisible by 2. That is, p^2 must be even.

However, if p^2 is even then p must also be even (since even $\times$ even = even, i.e. even^2 = even).

So we can write $p = 2n$

which gives $p^2 = 4n^2$

but previously we obtained

$$p^2 = 2q^2$$

hence $2q^2 = 4n^2$

and therefore $q^2 = 2n^2$

which means that q^2 is even. q must therefore also be even.

Since p and q are both even, $\dfrac{p}{q}$ cannot be in its simplest form. Since we have contradicted our original assumption it must have been false. Hence, we cannot assume $\dfrac{p}{q} = \sqrt{2}$.

So $\sqrt{2}$ cannot be a rational number and must therefore be irrational.

Appendix 2
Percentage

The percentage key, $\boxed{\%}$ (p.127)

This key works in different ways for different makes of calculator. For finding a percentage of something, as in Exercise 11 on p.127, it is probably easier to use the method given in the example and not to use the percentage key.

To work out $17\frac{1}{2}\%$ of £30 most calculators will probably give the correct answer if you key in:

$\boxed{1}\ \boxed{7}\ \boxed{\cdot}\ \boxed{5}\ \boxed{\times}\ \boxed{3}\ \boxed{0}\ \boxed{\%}$

Some calculators may need an equals sign:

$\boxed{1}\ \boxed{7}\ \boxed{\cdot}\ \boxed{5}\ \boxed{\times}\ \boxed{3}\ \boxed{0}\ \boxed{\%}\ \boxed{=}$

Some calculators will accept this order:

$\boxed{1}\ \boxed{7}\ \boxed{\cdot}\ \boxed{5}\ \boxed{\%}\ \boxed{\times}\ \boxed{3}\ \boxed{0}\ \boxed{=}$

Appendix 3
Bearings

Back bearing (reverse bearing) (p.326)

The bearing of P from Q is 075°, so
the bearing of Q from P is 255°.
(75° + 180° = 255°)
There are a number of ways of
reasoning why 180° should be
added in this example.

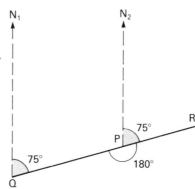

Reasoning 1

In the diagram, QP is produced to R and the two North lines have
been labelled N_1 and N_2.

$\angle N_1QP = \angle N_2PR = 75°$ (Corresponding angles: $QN_1//PN_2$.)

Required bearing $= \angle N_2PR + 180°$ (see the diagram)

$$= 075° + 180°$$

$$= \underline{\underline{255°}}$$

Reasoning 2

If you point from Q to P, then for the back bearing you must point
from P to Q (exactly the opposite direction).
To face the opposite direction you must turn through 180°.

So the bearing of 075°, for example, is increased by 180°.

∴   the back bearing $= 075° + 180°$

$$= \underline{\underline{255°}}$$

Reasoning 3

In the diagram, the North lines have been labelled N_1 and N_2 and line N_2P has been produced to S.

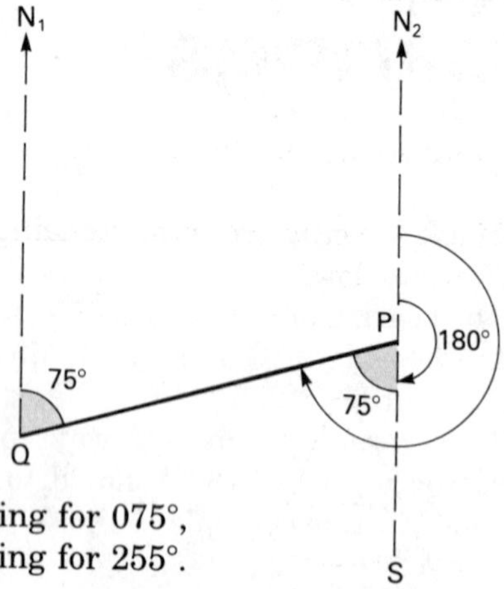

$\angle N_1QP = \angle SPQ = 75°$
(Alternate angles: $N_1Q /\!/ N_2S$.)

Required bearing $= 180° + \angle SPQ$

$$= 180° + 75°$$

$$= \underline{\underline{255°}}$$

Note Since 255° is the back bearing for 075°,
then 075° is the back bearing for 255°.

To find any back bearing from a given bearing, either add or subtract 180° to create an angle that lies between 0° and 360°.

Glossary

conic sections (p.32)

These are curves that are obtained from sections of a cone. They are shown below:

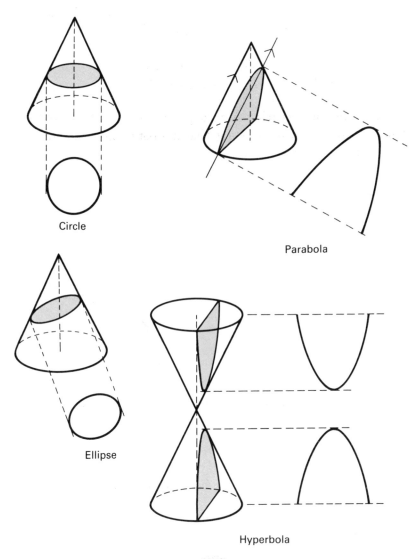

Circle

Parabola

Ellipse

Hyperbola

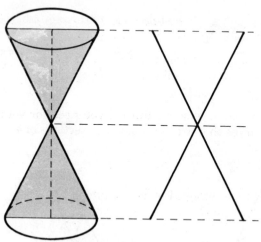

A pair of intersecting straight lines

cross-section   (p.166)

If you cut *across* a solid so that it is cut into two *sections*, the surface obtained by the cut is called the cross-section.

If you cut across a cylinder such that the cut is parallel to the base of the cylinder the cross-section obtained is a circle. The circle obtained is the same size as the circle at the base of the cylinder.

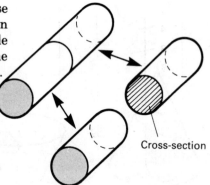

Cross-section

error   (p.11)

The correct term for the type of error used on p.11 is *absolute error*.

natural numbers   (p.1)

The natural numbers have long been accepted as being the set of counting numbers: $\{1, 2, 3, 4, 5, \dots\}$ which does not include zero. More recently, some bodies have included zero in the set of natural numbers.

Throughout this course, I have taken the natural numbers to be the counting numbers.

Note that some examining groups include zero as a natural number and therefore take whole numbers and natural numbers to be one and the same: $\{0, 1, 2, 3, 4, 5, \ldots\}$.

Make certain you know which definition you need for your course.

non-integral   (p.284)

A number that is non-integral is not an integer.

range   (p.203)

In a mapping, the first set is called the *domain* and the second set the *co-domain*. The set of elements that are actually used in the second set is called the *range*. Consider the two mapping diagrams below:

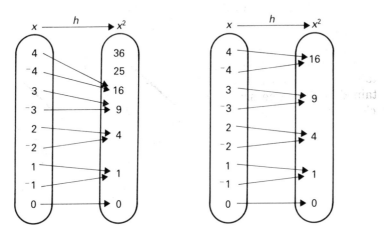

In both mappings the domain is $\{4, {}^-4, 3, {}^-3, 2, {}^-2, 1, {}^-1, 0\}$.

In both mappings the range is $\{16, 9, 4, 1, 0\}$

In the second mapping the co-domain is the same as the range. It is the set $\{16, 9, 4, 1, 0\}$. Note that all the elements are used.

In the first mapping the co-domain and the range are different. The co-domain is the set $\{36, 25, 16, 9, 4, 1, 0\}$ while the range $\{16, 9, 4, 1, 0\}$ is a proper subset of the co-domain.

sterling   (p.139)

Sterling is British money as distinguished from foreign money.

subset   (p.6)

If all the elements of set A are also members of set B then A is a subset of B.

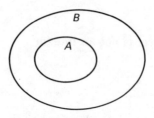